Second Edition

APPLIED CODEOLOGY

Navigating the NEC 2008

In Partnership with the NJATC

THOMSON

DELMAR LEARNING

Australia Canada Mexico Singapore Spain United Kingdom United States

Applied Codeology: Navigating the NEC 2008, Second Edition
NJATC

Vice President, Technology Professional Business Unit:
Gregory L. Clayton

Product Development Manager:
Ed Francis

Product Manager:
Ohlinger Publishing Services

Editorial Assistant:
Nobina Chakraborti

Director of Marketing:
Beth A. Lutz

Executive Marketing Manager:
Taryn Zlatin

Marketing Specialist:
Marissa Maiella

Director of Technology:
Paul Morris

Technology Project Manager:
Jim Ormsbee

Director of Production:
Patty Stephan

Production Manager:
Andrew Crouth

Content Project Manager:
Andrea Majot

Art Director:
Bethany Casey

Library of Congress Cataloging-in-Publication Data:
Card Number:
2007037922

ISBN-10: 1-4180-7349-0
ISBN-13: 978-1-4180-7349-7

NOTICE TO THE READER

CONTENTS

An electrician's apprentice, a journeyman wireman, an electrical engineer, or an electrical inspector will use the *National Electrical Code*® *(NEC)* daily. Their livelihood depends on their ability to properly install, design, and inspect electrical systems in accordance with the *NEC*®.

The *Applied Codeology* text and associated lessons are designed to teach you how to use the *NEC*®. This course is not "article" or "topic" specific. The *NEC*® may seem confusing. In addition, the *NEC*® changes every 3 years. A new apprentice may become overwhelmed with the amount of information in the *NEC*®. Many electrical workers complete an apprenticeship and engineers complete college, yet neither may feel comfortable using the *NEC*®.

The objective of this course is to give you confidence in your ability to quickly find pertinent information in the *NEC*®. To achieve this comfort level and genuine confidence in using the *NEC*®, read and apply the methods outlined in this text along with completing the associated homework lessons.

An experienced tradesman can differentiate between members of different trades just by looking at the tools they carry. Electricians use hand tools, the most identifiable being side-cutting pliers. To an apprentice, a pair of side-cutters seems clumsy, but necessary. After a few years of field experience, those same side-cutters seem to be an extension of the hand, capable of many tasks, all of which are performed quickly and efficiently. If you invest the time that this course requires, and focus on the methods described within, the *NEC*® will become a tool at your side that is just as comfortable and natural to use as your side-cutters.

ACKNOWLEDGMENTS

Principal writer Jim Dollard, IBEW Local Union 98, Philadelphia, PA,
NJATC technical writer and editor Palmer Hickman, NJATC Director

ADDITIONAL ACKNOWLEDGMENTS

This material is continually reviewed and evaluated by Training Directors
who are also members of the NJATC Inside Education Committee. The in-
valuable input provided by these individuals allows for the development of
instructional material that is of the absolute highest quality. At the time of
this printing, the Inside Education Committee was composed of the following members: Dennis Anthony (Phoenix, AZ); Byron Benton (San Leandro,
CA); John Biondi (Vineland, NJ); Bill Bowser (Renton, WA); Dan Campbell
(Tangent, OR); Eric Davis (Warren, OH); Lawrence Hidalgo (Lansing, MI); Greg
Hojdila (Beaver, PA); Chris Kelly (Hauppauge, NY); Tom Minder (Fairbanks,
AK); Jim Patterson (Indianapolis, IN); Janet Skipper (Winter Park, FL); Gary
Strouz (Houston, TX); Jim Sullivan (Winter Park, FL); Chris Thorsen (Evans-
ville, IN); Andrew White (White Plains, NY).

1

The Development
of the *National Electrical Code*®

OBJECTIVES

After completing this unit, you should be able to:
1. Explain the meaning of *consensus standard*
2. Understand that the *NEC*® is an ANSI-recognized consensus standard
3. Explain the proposal and comment process in the 3-year revision cycle of the *NEC*®
4. Identify the makeup of different classifications of a Code-Making Panel or Technical Committee
5. Explain how you can take part in the *NEC*® revision process

OVERVIEW

The *National Electrical Code (NEC*®) is revised every 3 years. This process is never ending. Before a new edition of the *NEC*® leaves the printer and is available to code users, proposals to change it are on file with the National Fire Protection Association (NFPA). Technical Committee members and NFPA staff also begin work on the next edition of the *NEC*® through task groups assigned to address issues identified in the previous cycle. These task groups work on technical issues and usability issues to formulate proposals for the next *NEC*® cycle. The *NEC*® is a living document, a work in progress that is shaped, molded, and improved by all members of the public who take part in the process.

THE *NEC®* PROCESS

The *National Electrical Code* (*NEC®*) is developed through a **consensus standards** development process approved by the American National Standards Institute (ANSI). This process includes input from all interested persons through the proposal and comment stages of the process. **Technical Committees** (or TCs), also known as Code-Making Panels (or CMPs), are formed to achieve consensus on all proposed changes or revisions to the *NEC®*. These committees are balanced to provide all viewpoints and interest groups a voice in the deliberations of issues brought before the committees. Technical Committee members are volunteers who represent a specific membership classification.

As a revision cycle comes to an end, the products are a new and improved *NEC®*, the **Report on Proposals (ROP)**, and the **Report on Comments (ROC)**. The Technical Committees work diligently to provide the user of the *NEC®* with code text that is easy to read and understand, practical, and enforceable. However, as changes in the *NEC®* are applied, users may disagree on the implementation of the revised requirements. The intent of the change can be found in the substantiations, CMP statements, and CMP member comments in both the ROP and ROC. All persons submitting a proposal receive a copy of the ROP in either paper, compact disc form or online. All persons submitting a comment to a proposal receive a copy of the ROC in paper, compact disc form or online. These documents are an extremely valuable tool for the user of the next edition of the *NEC®*. In the back of each edition of the *NEC®*, a "Proposal Form" is included. Anyone can use this form to submit a proposed change to the next edition of the *NEC®*. Proposals may also be submitted online at the National Fire Protection Association (NFPA) website, www.nfpa.org.

SEQUENCE OF EVENTS

Following is a discussion of the consensus standards development process.

The Proposal Stage

Proposals are submitted before the deadline given in the back of the *NEC®*. The proposals are then organized by the NFPA staff and forwarded to members of the 20 **Code-Making Panels**. The Code-Making Panels meet at the location and on the dates as stated in the back of the *NEC®* to act on all proposals. These meetings are open to all interested persons who want to observe the panel proceedings. Actions at the meeting require only a simple majority of votes; the actions taken by the Code-Making Panels in the proposal stage are as follows:

- Accept. The panel accepts the proposal as written; no panel statement is required.
- Reject. The panel rejects the proposal; a panel statement explaining why is required.
- Accept in Principle. The panel accepts the proposal in principle but changes the wording or takes other action to achieve the submitter's intention; a panel statement explaining why is required.

DidYouKnow?

A two-thirds majority on the written ballots is required to change the *NEC®*.

- Accept in Part. The panel accepts only part of the proposal and rejects the remainder; a panel statement explaining what part was accepted and why the remainder was rejected is required.
- Accept in Principle in Part. The panel accepts part of the proposal in principle but rejects the remainder; a panel statement is required.

After the panel meetings have closed, the written ballots are sent to all panel members. A two-thirds majority on the written ballot is required for the panel action to be upheld. The **Technical Correlating Committee** reviews all of the Code-Making Panel results to ensure that there are not conflicting actions. The ROP is printed and distributed (Figure 1–1).

FIGURE 1–1   Report on proposals.

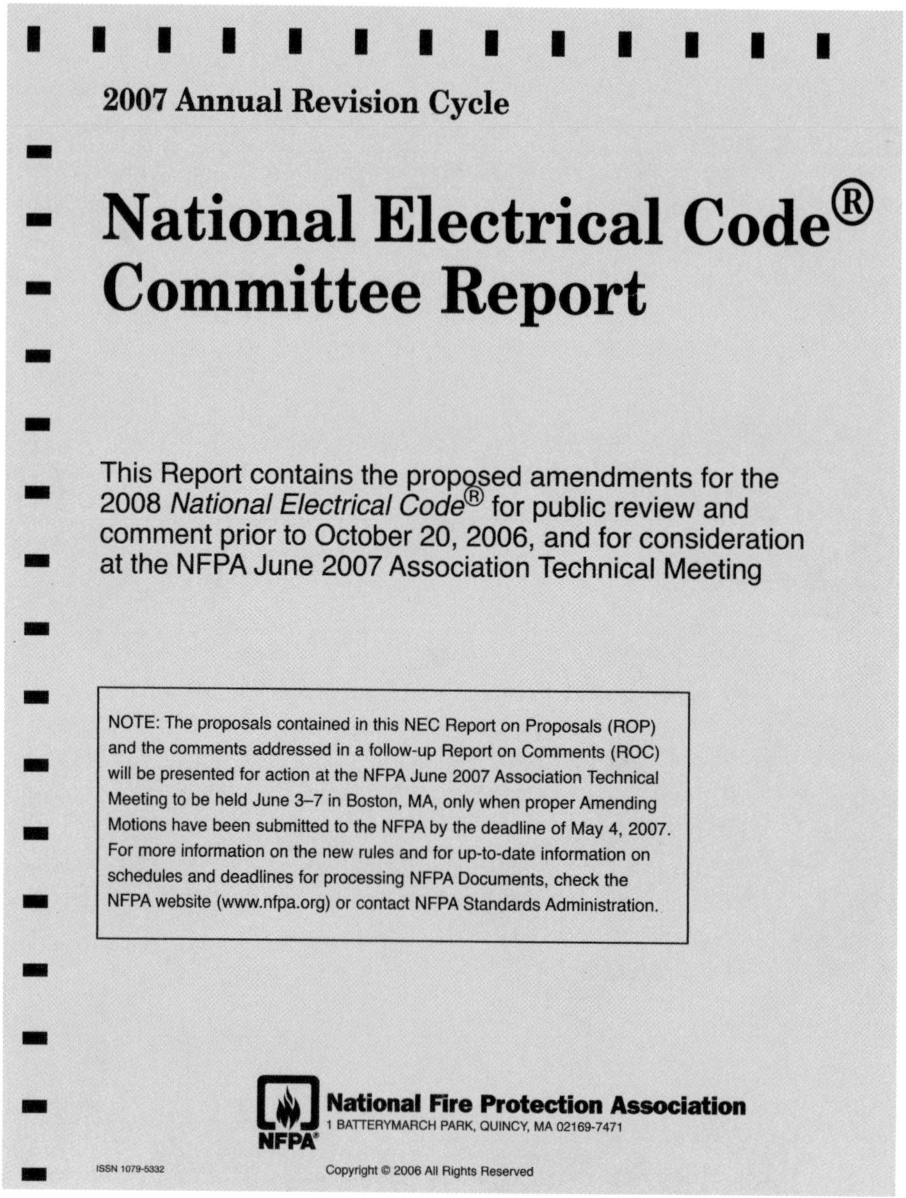

2007 Annual Revision Cycle

National Electrical Code®
Committee Report

This Report contains the proposed amendments for the 2008 *National Electrical Code*® for public review and comment prior to October 20, 2006, and for consideration at the NFPA June 2007 Association Technical Meeting

NOTE: The proposals contained in this NEC Report on Proposals (ROP) and the comments addressed in a follow-up Report on Comments (ROC) will be presented for action at the NFPA June 2007 Association Technical Meeting to be held June 3–7 in Boston, MA, only when proper Amending Motions have been submitted to the NFPA by the deadline of May 4, 2007. For more information on the new rules and for up-to-date information on schedules and deadlines for processing NFPA Documents, check the NFPA website (www.nfpa.org) or contact NFPA Standards Administration.

National Fire Protection Association
1 BATTERYMARCH PARK, QUINCY, MA 02169-7471
NFPA®

ISSN 1079-5332   Copyright © 2006 All Rights Reserved

The Comment Stage

Comments on the actions taken on the proposals are submitted before the deadline given in the back of the *NEC*®. The comments are then organized by the NFPA staff and forwarded to members of the 20 Code-Making Panels. The Code-Making Panels meet at the location and on the dates stated in the back of the *NEC*® to act on all comments. These meetings are open to all interested persons who want to observe the panel proceedings. Actions at the meeting require only a simple majority of votes; the actions that may be taken by the Code-Making Panels in the comment stage are the same as those in the proposal stage with one additional action permitted. Where a comment introduces new material that has not had public review or would require more time or work to process than allowed, an action of "Hold" is permitted. A panel statement explaining the action is required. A comment that is put on "Hold" returns in the next cycle as a proposal.

After the panel meetings have closed, the written ballots are sent to all panel members. A two-thirds majority on the written ballot is required for the panel action to be upheld. The Technical Correlating Committee reviews all of the Code-Making Panel results to ensure that there are not conflicting actions. The ROC is printed and distributed (Figure 1–2).

The NFPA Annual Meeting

The full *NEC*® technical committee report is presented to the NFPA membership for approval. Motions may be made to amend or reverse the actions taken by the Code-Making Panels. Any successful amendments made by the membership at the NFPA Annual Meeting are sent to the proper Code-Making Panel for a vote.

Appeals to the Standards Council

Appeals to the Standards Council may be made after the Annual Meeting. The Standards Council decides any appeals made, accepts the new *NEC*®, and issues the revised code.

COMMITTEE MEMBERSHIP CLASSIFICATIONS

The Code-Making Panels, known as the Technical Committees (TCs), are made up of volunteers. The committee list and scope for each Code-Making Panel is in the front of the *NEC*®. Each TC member has an identification letter/s listed with their name and employer. This identification is in brackets, such as L for Labor. The committee membership classification is part of the balancing process of each panel. Most organizations represented have a principal member and an alternate (see Figure 1-3). The committee classifications are as follows:

M	Manufacturers: makers of products affected by the *NEC*®
U	Users: users of the *NEC*®
I/M	Installers/Maintainers: installers/maintainers of systems covered by the *NEC*®
L	Labor: concerned with safety in the workplace

FIGURE 1–2   Report on comments.

2007 Annual Revision Cycle

National Electrical Code®
Committee Report
on Comments

NOTE: Notice of Intent to Make an
NEC® Motion (NITMAM) deadline is
May 4, 2007

A compilation of the documented action on
comments received by the code-making
panels for the 2007 Annual Revision Cycle

NOTE: The proposals contained in the NEC Report on Proposals (ROP)
and the comments addressed in this Report on Comments (ROC) will be
presented for action at the NFPA June 2007 Annual Association Technical
Meeting to be held June 3–7 in Boston, MA, only when proper Amending
Motions have been submitted to the NFPA by the deadline of May 4, 2007.
For more information on the new rules and for up-to-date information on
schedules and deadlines for processing NFPA Documents, check the NFPA
website (www.nfpa.org) or contact NFPA Standards Administration.

NFPA
1 BATTERYMARCH PARK, QUINCY, MA 02169-7471

ISSN 1079-5332   Copyright © 2007 All Rights Reserved

R/T   Research/Testing Labs: independent organizations developing/enforcing standards

E   Enforcing Authority: inspectors, enforcers of the *NEC*®

I   Insurance: insurance companies, bureaus, or agencies

C   Consumers: purchasers of products/systems not included in "U," users

SE   Special Experts: provide special expertise, not applicable to other classifications

UT   Utilities: installers/maintainers of systems not covered by *NEC*®

FIGURE 1-3 Each Code-Making Panel (CMP) is assigned a committee scope over which it has primary responsibility. For example, the scope of CMP-1 includes Articles 90, 100, 110, Annex A, and Annex G.

NEC® Code-Making Panel	Articles, Annex and Chapter 9 material within the scope of the Code-Making Panel
1	90, 100, 110, Annex A, Annex G
2	210, 215, 220, Annex D Examples 1 through 6
3	300, 590, 720, 725, 727, 760, Chapter 9 Tables 11(a) and (b), Tables 12(a) and (b)
4	225 & 230
5	200, 250, 280, 285
6	310, 400, 402, Chapter 9 Tables 5 through 9, Annex B
7	320, 322, 324, 326, 328, 330, 332, 334, 336, 338, 340, 382, 394, 396, 398
8	342, 344, 348, 350, 352, 353, 354, 355, 356, 358, 360, 362, 366, 368, 370, 372, 374, 376, 378, 380, 384, 386, 388, 390, 392, Chapter 9 Tables 1 through 4, Annex C
9	312, 314, 404, 408, 450, 490
10	240
11	409, 430, 440, 460, 470, Annex D, Example D8
12	610, 620, 625, 626, 630, 640, 645, 647, 650, 660, 665, 668, 669, 670, 685, Annex D, Examples D9 and D10
13	445, 455, 480, 690, 692, 695, 700, 701, 702, 705
14	500, 501, 502, 503, 504, 505, 506, 510, 511, 513, 514, 515, 516
15	517, 518, 520, 522, 525, 530, 540
16	770, 800, 810, 820, 830
17	422, 424, 426, 427, 680, 682
18	406, 410, 411, 600, 605
19	545, 547, 550, 551, 552, 553, 555, 604, 675, Annex D, Examples D11 and D12
20	708

Many organizations take part in the *NEC*® process and provide representation for the membership classification that applies to the organization. For example, the members of the Technical Committees, with the classification E for Enforcing Authority, are representatives of the IAEI, International Association of Electrical Inspectors. The organizations represented on the Technical Committees are as follows:

Air Conditioning and Refrigeration Institute

Alliance for Telecommunications Industry Solutions

The Aluminum Association

American Chemistry Council

American Iron and Steel Institute

American Lighting Association

FIGURE 1–4 Committee list.

NATIONAL ELECTRICAL CODE COMMITTEE

Randall R. McCarver, Telcordia Technologies, Incorporated, NJ [U]
Rep. Alliance for Telecommunications Industry Solutions
Lanny G. McMahill, City of Phoenix, AZ [E]
Rep. International Association of Electrical Inspectors
H. Brooke Stauffer, National Electrical Contractors Association, MD [IM]
Rep. National Electrical Contractors Association
John W. Troglia, Edison Electric Institute, WI [UT]
Rep. Electric Light & Power Group/EEI

Alternates

Lawrence S. Ayer, Biz Com Electric, Incorporated, OH [IM]
Rep. Independent Electrical Contractors, Incorporated
(Alt. to D. L. Hittinger)
Kenneth P. Boyce, Underwriters Laboratories Incorporated, IL [RT]
(Alt. to D. A. Dini)
Ernest J. Gallo, Telcordia Technologies, Incorporated, NJ [U]
Rep. Alliance for Telecommunications Industry Solutions
(Alt. to R. R. McCarver)
Russell J. Helmick, Jr., City of Irvine, CA [E]
Rep. International Association of Electrical Inspectors
(Alt. to L. G. McMahill)

Neil F. LaBrake, Jr., Niagara Mohawk, a National Grid Company, NY [UT]
Rep. Electric Light & Power Group/EEI
(Alt. to J. W. Troglia)
Donald H. McCullough, II, Westinghouse Savannah River Company, SC [U]
Rep. Institute of Electrical & Electronics Engineers, Incorporated
(Alt. to H. L. Floyd II)
Gil Moniz, National Electrical Manufacturers Association, MA [M]
Rep. National Electrical Manufacturers Association
(Alt. to J. D. Minick)
Rick Munch, Frischhertz Electric Company, LA [L]
Rep. International Brotherhood of Electrical Workers
(Alt. to P. L. Hickman)

Nonvoting

Ark Tsisserev, City of Vancouver, BC, Canada
Rep. Canadian Standards Association International

CODE–MAKING PANEL NO. 2
Articles 210, 215, 220, Annex D, Examples 1 through 6

Raymond W. Weber, *Chair*
State of Wisconsin, WI [E]
Rep. International Association of Electrical Inspectors

Richard W. Becker, Engineered Electrical Systems, Incorporated, WA [U]
Rep. Institute of Electrical & Electronics Engineers, Incorporated
Frank Coluccio, New York City Department of Buildings, NY [E]
Matthew D. Dobson, National Association of Home Builders, DC [U]
Rep. National Association of Home Builders
Thomas L. Harman, University of Houston/Clear Lake, TX [SE]
Donald M. King, IBEW Local Union 313, DE [L]
Rep. International Brotherhood of Electrical Workers
Christopher P. O'Neil, National Grid USA Service Company, MA
Rep. Electric Light & Power Group/EEI, MA
James T. Pauley, Square D Company, KY [M]
Rep. National Electrical Manufacturers Association
Susan W. Porter, Underwriters Laboratories Incorporated, NY [RT]
Joseph Patterson Roché, Celanese Acetate, SC [U]
Rep. American Chemistry Council
Albert F. Sidhom, U.S. Army Corps of Engineers, CA [U]
Michael D. Toman, MEGA Power Electrical Services, Incorporated, MD [IM]
Rep. National Electrical Contractors Association
Robert G. Wilkinson, Independent Electrical Contractors of Texas Gulf Coast, TX [IM]
Rep. Independent Electrical Contractors, Incorporated

Alternates

Kevin J. Brooks, IBEW Local Union 16, IN [L]
Rep. International Brotherhood of Electrical Workers
(Alt. to D. M King)

Ernest S. Broome, City of Knoxville, TN [E]
Rep. International Association of Electrical Inspectors
(Alt. to R. W. Weber)
James R. Jones, University of Alabama at Birmingham, AL [U]
Rep. Institute of Electrical & Electronics Engineers, Incorporated
(Alt. to R. W. Becker)
Daniel J. Kissane, Pass & Seymour/Legrand, NY [M]
Rep. National Electrical Manufacturers Association
(Alt. to J. T. Pauley)
Brian J. Nenninger, The Dow Chemical Company, TX [U]
Rep. American Chemistry Council
(Alt. to J. P. Roché)
Clifford L. Rediger, Independent Electrical Contractors Training Fund, CO [IM]
Rep. Independent Electrical Contractors, Incorporated
(Alt. to R. G. Wilkinson)
Richard V. Wagner, Underwriters Laboratories Incorporated, NY [RT]
(Alt. to S. W. Porter)
Joseph E. Wiehagen, National Association of Home Builders, MD [U]
Rep. National Association of Home Builders
(Alt. to M. D. Dobson)

Nonvoting

Douglas A. Lee, U.S. Consumer Product Safety Commission, MD [C]
Andrew M. Trotta, U.S. Consumer Product Safety Commission, MD [C]
(Alt. to D. A. Lee)

NATIONAL ELECTRICAL CODE COMMITTEE

CODE–MAKING PANEL NO. 3
Articles 300, 590, 720, 725, 727, 760, Chapter 9, Tables 11(a) and (b), and Tables 12(a) and (b)

Richard P. Owen, *Chair*
City of St. Paul, MN [E]
Rep. International Association of Electrical Inspectors

Lawrence S. Ayer, Biz Com Electric, Incorporated, OH [IM]
　Rep. Independent Electrical Contractors, Incorporated
Paul J. Casparro, Scranton Electricians JATC, PA [L]
　Rep. International Brotherhood of Electrical Workers
Les Easter, Allied Tube and Conduit, IL [M]
　Rep. National Electrical Manufacturers Association
Sanford E. Egesdal, Egesdal Associates PLC, MN [M]
　Rep. Automatic Fire Alarm Association, Incorporated
Thomas J. Guida, Underwriters Laboratories, Inc., NY [RT]
Dennis B. Horman, PacifiCorp, UT [UT]
　Rep. Electric Light & Power Group/EEI
Ray R. Keden, ERICO, Incorporated, CA [M]
　Rep. Building Industry Consulting Services International
Ronald E. Maassen, Lemberg Electric Company, Incorporated, WI [IM]
　Rep. National Electrical Contractors Association
Steven J. Owen, Steven J. Owen, Incorporated, AL [IM]
　Rep. Associated Builders and Contractors, Incorporated
David A. Pace, Olin Corporation, AL [U]
　Rep. American Chemistry Council
Melvin K. Sanders, Things Electrical Company, Incorporated (TECo., Incorporated), IA [U]
　Rep. Institute of Electrical & Electronics Engineers, Incorporated
John E. Sleights, Travelers Insurance, CT [I]

Alternates
Mark E Christian, Chattanooga Electrical JATC, TN [L]
　Rep. International Brotherhood of Electrical Workers
　(Alt. to P. J. Casparro)
Dr. Shane M. Clary, Bay Alarm Company, Incorporated, CA [M]
　Rep. Automatic Fire Alarm Association, Incorporated
　(Alt. to S. E. Egesdal)

Adam D. Corbin, Corbin Electrical Services, Incorporated, NJ [IM]
　Rep. Independent Electrical Contractors, Inc.
　(Alt. to L. S. Ayer)
John C. Hudak, Old Forge, PA [E]
　Rep. International Association of Electrical Inspectors
　(Alt. to R. P. Owen)
Danny Liggett, DuPont Engineering, DE [U]
　Rep. American Chemistry Council
　(Alt. to D. A. Pace)
Juan C. Menendez, Southern California Edison Company, CA [UT]
　Rep. Electric Light & Power Group/EEI
　(Alt. to D. B. Horman)
T. David Mills, Bechtel Savannah River, Incorporated, SC [U]
　Rep. Institute of Electrical & Electronics Engineers, Incorporated
　(Alt. to M. K. Sanders)
Mark C. Ode, Underwriters Laboratories Incorporated, NC [RT]
　(Alt. to T. J. Guida)
Lorena Orbanic, Carlon, Lamson & Sessions, OH [M]
　Rep. Building Industry Consulting Services International
　(Alt. to R. R. Keden)
Roger S. Passmore, Davis Electrical Constructors, Incorporated, SC [IM]
　Rep. Associated Builders and Contractors, Incorporated
　(Alt. to S. J. Owen)
George A. Straniero, AFC Cable Systems, Incorporated, NJ [M]
　Rep. National Electrical Manufacturers Association
　(Alt. to L. Easter)

CODE–MAKING PANEL NO. 4
Articles 225, 230

James M. Naughton, *Chair*
IBEW Local Union 103, MA [L]
Rep. International Brotherhood of Electrical Workers

Malcolm Allison, Ferraz Shawmut, MA [M]
C. John Beck, Pacific Gas and Electric Company, CA [UT]
　Rep. Electric Light & Power Group/EEI
Robert J. Deaton, The Dow Chemical Company, TX [U]
　Rep. Institute of Electrical & Electronics Engineers, Incorporated
Howard D. Hughes, Hughes Electric Company Incorporated, AR [IM]
　Rep. National Electrical Contractors Association
William M. Lewis, Eli Lilly & Company, IN [U]
　Rep. American Chemistry Council
Mark C. Ode, Underwriters Laboratories Incorporated, NC [RT]
James J. Rogers, Towns of Oak Bluffs, Tisbury, West Tisbury, MA [E]
　Rep. International Association of Electrical Inspectors

John W. Young, Siemens Energy & Automation, Incorporated, GA [M]
　Rep. National Electrical Manufacturers Association
Vincent Zinnante, Advantage Electric, Incorporated, TX [IM]
　Rep. Independent Electrical Contractors, Incorporated

Alternates
Thomas L. Adams, Exelon Corporation, IL [UT]
　Rep. Electric Light & Power Group/EEI
　(Alt. to C. J. Beck)
Ronald Breschini, Underwriters Laboratories Incorporated, CA [RT]
　(Alt. to M. C. Ode)
Terry D. Cole, Hamer Electric, WA [IM]
　Rep. Independent Electrical Contractors, Incorporated
　(Alt. to V. Zinnante)

DidYouKnow?

Each code-making panel consists of a chairman, principal members, and alternate members.

DidYouKnow?

Only principal members may vote in the *NEC*® process. The alternate's vote is only counted when the principal member does not return the written ballot.

American Petroleum Institute

American Society of Agricultural Engineers

American Society of Anesthesiologists

American Society for Healthcare Engineering

Associated Builders and Contractors

Association of Higher Education Facilities Officers

Association of Home Appliance Manufacturers

Association of Iron and Steel Engineers

Building Industry Consulting Service International

Canadian Standards Association International

Copper Development Association Incorporated

Edison Electrical Institute

Electrical Generating Systems Association

Illuminating Engineering Society of North America

Independent Electrical Contractors

Industrial Risk Insurers

Information Technology Industry Council

Institute of Electrical and Electronics Engineers

Insulated Cable Engineers Association Incorporated

International Association of Electrical Inspectors

International Association of Theatrical Stage Employees

International Brotherhood of Electrical Workers

International Electrical Testing Association Incorporated

International Society for Measurement and Control

Manufactured Housing Institute

Motion Picture Association of America Incorporated

National Association of Home Builders

National Association of RV Parks and Campgrounds

National Cable Television Association

National Electrical Contractors Association

National Electrical Manufacturers Association

National Elevator Incorporated

National MultiHousing Council

National Spa and Pool Institute

New York Board of Fire Underwriters

Outdoor Amusement Business Association Incorporated

Power Tool Institute Incorporated

Recreation Vehicle Industry Association

Society of Automotive Engineers

Society of the Plastics Industry Incorporated

Solar Energy Industries Association

U.S. Fuel Cell Council

U.S. Institute for Theatre Technology

SUMMARY

The *NEC®* revision process is open to all members of the public who wish to take part by submitting proposed changes. The many organizations that take part in the *NEC®* code-making process help to build a true consensus code. The *NEC®* will change every three years; memorizing requirements or sections may be a wasted effort due to the three-year revision cycle of the *NEC®*. The codeology method, however, will not change. Applying the codeology method will lead to the quick and accurate location of necessary information in the *NEC®* today and in all future and past versions of the code.

REVIEW QUESTIONS

1. Proposals and comments sent to the National Fire Protection Association to change NFPA-70, the National Electrical Code, may be submitted by _____.
   a. organizations represented in the code-making process
   b. large manufacturers of electrical equipment
   c. anyone who is interested
   d. NFPA members

2. Actions taken by the Technical Committee or Code-Making Panel require a _____ vote at the ROP or ROC meeting for passage.
   a. written
   b. chairman's
   c. two-thirds
   d. simple majority

3. Actions taken by the Technical Committee or Code-Making Panel require a _____ vote on the written ballot of the ROP or ROC for passage.
   a. written
   b. chairman's
   c. two-thirds
   d. simple majority

4. The *National Electrical Code* is an ANSI document, which means it is developed through a _____ standards development process.
   a. consensus
   b. private
   c. members-only
   d. governmental

5. The committee membership classification "U" designates a person representing _____.
   a. utilities
   b. users
   c. underwriters
   d. United States government

6. When a comment is submitted containing new material which did not have public review in the ROP, the action taken by the CMP will be to _____.
   a. Accept
   b. Reject
   c. Accept in Principle
   d. Hold

7. A panel statement is required on all actions taken in the proposal and comment stages except for an action to _____.
   a. Reject
   b. Accept
   c. Accept in Part
   d. Hold

8. The first step in the *NEC®* revision process is the _____.
   a. proposal stage
   b. comment stage
   c. NFPA annual meeting
   d. appeals to the Standards Council

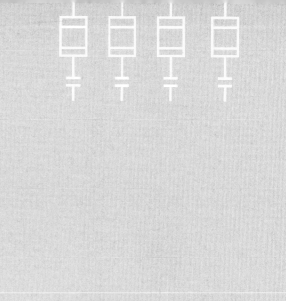

2

Basic Building Block #1: Table of Contents

NEC® TABLE OF CONTENTS
The Structure of the Table of Contents

OBJECTIVES

After completing this unit, you should be able to:
1. Recognize the importance of being extremely familiar with the *NEC*® Table of Contents
2. Explain why the table of contents is the starting point for all inquiries in the *NEC*® using the Codeology method
3. Recognize the ten major subdivisions of the *NEC*®
4. Recognize and understand the numbering method of articles within the chapters of the *NEC*®

OVERVIEW

This unit will introduce the first of four basic building blocks that are essential for the understanding and application of the Codeology method. Throughout this text, these basic principles will be repeated time and time again until they become part of our natural thought process when working with the *NEC*®.

When you open the *NEC*® or any other type of installation code, it will begin with a table of contents placed in the front of the document, by design, to quickly take the user to the information he/she desires.

NEC® TABLE OF CONTENTS

The **table of contents** in the *NEC*® is the starting point for all of our inquiries in the *National Electrical Code*®. Once a need or question arises for information contained in the *NEC*®, the next step is to use the table of contents to get into the right **chapter, article,** and **part.** It is essential that all users of the *NEC*® be extremely familiar with the contents of the *NEC*®.

The table of contents is broken down into ten separate pieces. The first is the Introduction to the *NEC*®, Article 90. The *NEC*® is then subdivided into nine chapters that are major subdivisions of the *NEC*® covering broad areas. These chapters are then subdivided into articles.

The table of contents is broken down into ten separate pieces as follows:

Article 90	Introduction
Chapter 1	General
Chapter 2	Wiring and Protection
Chapter 3	Wiring Methods and Materials
Chapter 4	Equipment for General Use
Chapter 5	Special Occupancies
Chapter 6	Special Equipment
Chapter 7	Special Conditions
Chapter 8	Communications Systems
Chapter 9	Tables & Annexes

The Structure of the Table of Contents

A basic understanding of these ten major subdivisions of the *NEC*® is essential to the **Codeology** method and must be committed to memory. The Introduction and each chapter will be covered completely in other chapters of the Codeology text. The objective of this unit is to introduce and explain the structure of the table of contents as follows:

Article 90, Introduction

The first major subdivision of the *NEC*® is Article 90, the Introduction, which provides basic information and requirements necessary to properly apply the rest of the document. Article 90 provides the user with the ground rules upon which the rest of the *NEC*® is written. In laying the ground rules, Article 90 details, for example:

Sections	90.1 Purpose
	90.2 Scope
	90.3 Code Arrangement

Chapter 1, General

Provisions that apply generally to all electrical installations are contained in Chapter 1. The articles in the 100 series are as follows:

Article 100 Definitions

Article 110 Requirements for Electrical Installations

DidYouKnow?

The table of contents is always the starting point when using the *NEC*®.

DidYouKnow?

The nine chapters of the *NEC*® logically separate installation requirements to aid the code user to quickly locate needed information.

DidYouKnow?

General requirements in Chapter 1 include definitions used in more than one article.

FIGURE 2–1   *NEC®* table of contents.

CONTENTS

Contents

CONTENTS

DidYouKnow?

The broad scope of Chapter 2 is addressed as follows: *Wiring* in Articles 200, 210, 215, 220, 225, and 230 *Protection* in Articles 240, 250, 280, and 285.

Chapter 2, Wiring and Protection

Provisions for wiring and protection that apply generally to all electrical installations are contained in Chapter 2. The articles in the 200 series are as follows:

Article 200 Use and Identification of Grounded Conductors

Article 210 Branch Circuits

Article 215 Feeders

Article 220 Branch Circuit, Feeder, and Service Calculations

Article 225 Outside Branch Circuits and Feeders

Article 230 Services

Article 240 Overcurrent Protection

Article 250 Grounding and Bonding

Article 280 Surge Arresters, over 1 kV

Article 285 Surge-Protective Devices (SPDs), 1 kV or Less

Chapter 3, Wiring Methods and Materials

Provisions for all wiring methods and materials that apply generally to all electrical installations are contained in Chapter 3. The articles in the 300 series are as follows:

Article 300 Wiring Methods

Article 310 Conductors for General Wiring

Article 312 Cabinets, Cutout Boxes, and Meter Socket Enclosures

Article 314 Outlet, Device, Pull, and Junction Boxes; Conduit Bodies; Fittings; and Handhole Enclosures

Article 320 Armored Cable: Type AC

Article 322 Flat Cable Assemblies: Type FC

Article 324 Flat Conductor Cable: Type FCC

Article 326 Integrated Gas Spacer Cable: Type IGS

Article 328 Medium Voltage Cable: Type MV

Article 330 Metal Clad Cable: Type MC

Article 332 Mineral Insulated, Metal Sheathed Cable: Type MI

Article 334 Nonmetallic Sheathed Cable: Types NM, NMC, and NMS

Article 336 Power and Control Tray Cable: Type TC

Article 338 Service Entrance Cable: Types SE and USE

Article 340 Underground Feeder and Branch Circuit Cable: Type UF

Article 342 Intermediate Metal Conduit: Type IMC

Article 344 Rigid Metal Conduit: Type RMC

Article 348 Flexible Metal Conduit: Type FMC

Article 350 Liquidtight Flexible Metal Conduit: Type LFMC

Article 352 Rigid Nonmetallic Conduit: Type RNC

Article 353 High Density Polyethylene Conduit: Type HDPE Conduit

Article 354 Nonmetallic Underground Conduit with Conductors: Type NUCC

Article 355 Reinforced Thermosetting Resin Conduit: Type RTRC

Article 356 Liquidtight Flexible Nonmetallic Conduit: Type LFNC

DidYouKnow?

Article 300 is titled "Wiring Methods" and contains general information for the installation of all *wiring methods and materials*.

Chapter 4, Equipment for General Use

Provisions for all electrical equipment that apply generally to all electrical installations are contained in Chapter 4. The articles in the 400 series are as follows:

DidYouKnow?

Chapter 4 includes general requirements for equipment which utilizes electrical energy and equipment/materials necessary to facilitate this use.

Article 480 Storage Batteries

Article 490 Equipment Over 600 Volts, Nominal

Chapter 5, Special Occupancies

Provisions for supplementing or modifying the general rules in Chapters 1 through 4 for "Special Occupancies" are contained in Chapter 5. The articles in the 500 series are as follows:

Article 500 Hazardous (Classified) Locations, Classes I, II, and III, Divisions 1 and 2

Article 501 Class I Locations

Article 502 Class II Locations

Article 503 Class III Locations

Article 504 Intrinsically Safe Systems

Article 505 Class I, Zone 0, 1, and 2 Locations

Article 506 Zone 20, 21 and 22 Locations for Combustible Dusts, Fibers and Flyings

Article 510 Hazardous (Classified) Locations, Specific

Article 511 Commercial Garages, Repair and Storage

Article 513 Aircraft Hangars

Article 514 Motor Fuel Dispensing Facilities

Article 515 Bulk Storage Plants

Article 516 Spray Application, Dipping and Coating Processes

Article 517 Health Care Facilities

Article 518 Assembly Occupancies

Article 520 Theatres, Audience Areas of Motion Picture and Television Studios, Performance Areas, and Similar Locations

Article 522 Control Systems for Permanent Amusement Attractions

Article 525 Carnivals, Circuses, Fairs and Similar Events

Article 530 Motion Picture and Television Studios and Similar Locations

Article 540 Motion Picture Projection Rooms

Article 545 Manufactured Buildings

Article 547 Agricultural Buildings

Article 550 Mobile Homes, Manufactured Homes, and Mobile Home Parks

Article 551 Recreational Vehicles and Recreational Vehicle Parks

Article 552 Park Trailers

Article 553 Floating Buildings

Article 555 Marinas and Boatyards

Article 590 Temporary Installations

Chapter 6, Special Equipment

Provisions for supplementing or modifying the general rules in Chapters 1 through 4 for "Special Equipment" are contained in Chapter 6. The articles in the 600 series are as follows:

Article 600 Electric Signs and Outline Lighting

Article 604 Manufactured Wiring Systems

DidYouKnow?

Chapter 5 modifies and supplements the general rules of Chapters 1 through 4 for special occupancies.

DidYouKnow?

Chapter 6 modifies and supplements the general rules of Chapters 1 through 4 for special equipment.

Chapter 7, Special Conditions

Provisions for supplementing or modifying the general rules in Chapters 1 through 4 for "Special Conditions" are contained in Chapter 7. The articles in the 700 series are as follows:

Chapter 8, Communications Systems

Provisions for "Communications Systems" are contained in Chapter 8. The requirements of Chapters 1 through 7 apply to Chapter 8 articles only when

DidYouKnow?

Chapter 7 modifies and supplements the general rules of Chapter 1 through 4 for special conditions.

DidYouKnow?

Chapter 8 is independent of Chapters 1 through 7. Only requirements referenced in Chapter 8 will apply.

DidYouKnow?

Tables located in Chapter 9 apply as referenced elsewhere in the *NEC®*.

they are specifically referenced in a Chapter 8 article. The articles in the 800 series are as follows:

Article 800 Communications Circuits

Article 810 Radio and Television Equipment

Article 820 Community Antenna Television and Radio Distribution Systems

Article 830 Network Powered Broadband Communications Systems

Chapter 9, Tables

This chapter contains **tables** that are referenced in other chapters of the *NEC®*. **Annexes** are provided for informational purposes only.

Table 1 Percent of Cross Section of Conduit and Tubing for Conductors, *Conduit Fill.*

Table 2 Radius of Conduit and Tubing Bends

Table 4 Dimensions and Percent Area of Conduit and Tubing (Areas of Conduit or Tubing for the Combinations of Wires Permitted in Table 1, Chapter 9)

Table 5 Dimensions of Insulated Conductors and Fixture Wires

Table 5A Compact Aluminum Building Wire Nominal Dimensions and Areas

Table 8 Conductor Properties

Table 9 Alternating Current Resistance and Reactance for 600-Volt Cables, 3-Phase, 60 Hz, 75°C (167°F)-Three Single Conductors in Conduit

Table 11A Class 2 & Class 3 Alternating Current Power Source Limitations

Table 11B Class 2 & Class 3 Direct Current Power Source Limitations

Table 12A PLFA Alternating Current Power Source Limitations

Table 12B PLFA Direct Current Power Source Limitations

Annex A Product Safety Standards

Annex B Application Information for Ampacity Calculation

Annex C Conduit and Tubing Fill Tables for Conductors and Fixture Wires of the Same Size

Annex D Examples

Annex E Types of Construction

Annex F Availability and Reliability for Critical Operations Power Systems; and Development and Implementation of Functional Performance Tests (FPTs) for Critical Operations Power Systems

Annex G Supervisory Control and Data Acquisition (SCADA)

Annex H Administration and Enforcement

SUMMARY

The starting point for all inquiries in the *NEC®* is the table of contents. The *NEC®* is subdivided into ten major subdivisions as follows:

Article 90 Introduction
Chapter 1 General
Chapter 2 Wiring and Protection
Chapter 3 Wiring Methods and Materials
Chapter 4 Equipment for General Use

Chapter 5 Special Occupancies
Chapter 6 Special Equipment
Chapter 7 Special Conditions
Chapter 8 Communications Systems
Chapter 9 Tables & Annexes

Chapters are subdivided into articles which logically pertain to the broad scope of the individual chapter.

REVIEW QUESTIONS

1. The table of contents is broken down into _____ major subdivisions.
   a. ten
   b. eight
   c. two
   d. nine

2. The Introduction to the *National Electrical Code®* is located in _____ .
   a. Chapter 1
   b. The table of contents
   c. Chapter 2
   d. Article 90

3. The starting point for all inquiries into the *National Electrical Code®* using the Codeology method is _____ .
   a. your best guess
   b. the index
   c. the table of contents
   d. The *NEC®* handbook

4. Articles within the *National Electrical Code®* in the 500 series are dedicated in scope to _____ _____ .
   a. utilization equipment
   b. special occupancies
   c. special conditions
   d. wiring methods

5. Which chapter will contain an article to address the special equipment requirements for swimming pools?
   a. 8
   b. 7
   c. 6
   d. 5

6. The answer to a general question about the meaning of the term "ampacity" will be found in Chapter ___.
   a. 1
   b. 2
   c. 3
   d. 4

7. The answer to a question about the size of conductors supplying a motor in a general application will be found in Chapter ___.
   a. 1
   b. 2
   c. 3
   d. 4

8. The answer to a question about the application of a legally required standby system will be found in Chapter ___.
   a. 8
   b. 7
   c. 6
   d. 5

3

Basic Building Block #2: Section 90.3

O U T L I N E

OBJECTIVES

After completing this unit, you should be able to:
1. Recognize that the *NEC®* is organized in an outline form as required in Section 90.3
2. Determine that Article 90 contains introductory information and requirements
3. Recognize that Chapters 1 through 4 apply generally to all electrical installations
4. Recognize that Chapters 5, 6, and 7 are Special and will supplement or modify Chapters 1 through 4
5. Recognize that Chapter 8 stands alone; Chapters 1 through 7 will apply to Chapter 8 installations only when referenced in Chapter 8
6. Recognize that the tables in Chapter 9 apply as referenced in the *NEC®*
7. Recognize that the Annexes provided are informational only

OVERVIEW

This unit will introduce the second of four basic building blocks that are essential for the understanding and application of the Codeology method. Throughout this text, these basic principles will be repeated time and time again until they become part of our natural thought process when working with the *NEC®*.

When you open the *NEC®* or any other type of installation code, it is absolutely necessary to understand the arrangement and application of the requirements in that code. The arrangement of the *NEC®* is outlined in the Introduction to the *NEC®*, Article 90, in Section 90.3.

THE ARRANGEMENT OF THE *NEC®*, SECTION 90.3

One of the most basic yet extremely important facts about the structure of the *NEC®* is that the arrangement of the table of contents is specifically designed to facilitate the proper application of each chapter. In Article 90, the "Introduction," the *NEC®* details this **code arrangement** in Section 90.3. Section 90.3 describes the division of the *NEC®* into the Introduction and nine chapters, as shown in Table 3–1.

As required in Section 90.3, Chapters 1, 2, 3, and 4 will apply generally in all electrical installations. These first four chapters contain the basic electrical installation requirements for all electrical installations. This information applies generally to all electrical installations from a single-family dwelling unit to a petroleum refinery or a hospital. A comprehensive understanding of these four chapters is imperative because they are the backbone for all electrical installations.

Chapter 1—General

Although the title of Chapter 1 is simply "General," the scope of the chapter is *General* Information and Rules for Electrical Installations. The chapter contains two articles:

    100 Definitions
    110 Requirements for Electrical Installations

DidYouKnow?

Chapters 1 through 4 provide general installation requirements, which are the backbone of all electrical installations.

TABLE 3–1	Summary of the Arrangement of the *NEC®* in Accordance with Section 90.3

Chapters 1 through 4 apply GENERALLY to ALL electrical installations.	
Chapter 1	GENERAL
Chapter 2	WIRING and PROTECTION
Chapter 3	WIRING METHODS and MATERIALS
Chapter 4	EQUIPMENT for GENERAL USE
Chapters 5, 6, and 7 SUPPLEMENT or MODIFY Chapters 1 through 4.	
Chapter 5	SPECIAL OCCUPANCIES
Chapter 6	SPECIAL EQUIPMENT
Chapter 7	SPECIAL CONDITIONS
Chapters 1 through 7 DO NOT apply to Chapter 8 unless there is a specific reference in Chapter 8 referring to another chapter.	
Chapter 8	COMMUNICATIONS SYSTEMS
The tables in Chapter 9 apply as referenced elsewhere in the *NEC®*.	
Chapter 9	TABLES
Annexes A through H are for informational purposes only and are not mandatory.	
Chapter 9	ANNEXES A through H

DidYouKnow?

An in-depth understanding of the arrangement of the *NEC®* is essential for the proper application of electrical installation requirements.

Chapter 2—Wiring and Protection

Chapter 2's title is Wiring and Protection; however, the scope of this chapter is Information and Rules on Wiring and Protection of Electrical Installations. Articles addressing WIRING include information related to grounded conductors, calculations for conductor size, branch circuits, feeders and services. Articles addressing PROTECTION include information related to the use of overcurrent protection, grounding, bonding, surge arrestors and SPDs.

This chapter has 10 articles:

200 Use and Identification of Grounded Conductors

210 Branch Circuits

215 Feeders

220 Branch Circuit, Feeder, and Service Calculations

225 Outside Branch Circuit and Feeders

230 Services

240 Overcurrent Protection

250 Grounding and Bonding

280 Surge Arresters, over 1 kV

285 Surge-Protective Devices (SPDs), 1 kV or Less

Chapter 3—Wiring Methods and Materials

The scope of this chapter is Information and Rules on Wiring Methods and Materials for Use in Electrical Installations. The articles include information on all permitted methods and materials to supply an electrical installation. This chapter details requirements for all wiring methods and materials from the **service point** to termination at the last outlet in the electrical distribution system. It includes, for example, general information for all wiring methods, types of cable assemblies, types of raceways, cabinets, cutout boxes, meter socket enclosures, types of boxes, conduit bodies, fittings, and more.

This chapter contains 43 articles:

300 Wiring Methods

310 Conductors for General Wiring

312 Cabinets, Cutout Boxes, and Meter Socket Enclosures

314 Outlet, Device, Pull, and Junction Boxes; Conduit Bodies; Fittings; and Handhole Enclosures

320 Armored Cable: Type AC

322 Flat Cable Assemblies: Type FC

324 Flat Conductor Cable: Type FCC

326 Integrated Gas Spacer Cable: Type IGS

328 Medium Voltage Cable: Type MV

330 Metal Clad Cable: Type MC

332 Mineral Insulated, Metal Sheathed Cable: Type MI

334 Nonmetallic Sheathed Cable: Types NM, NMC, and NMS

336 Power and Control Tray Cable: Type TC

338 Service Entrance Cable: Types SE and USE

DidYouKnow?

All cable assembly and circular raceway Articles share a common article layout and section numbering system to provide the code user with a consistent, easy-to-use format.

340 Underground Feeder and Branch Circuit Cable: Type UF

342 Intermediate Metal Conduit: Type IMC

344 Rigid Metal Conduit: Type RMC

348 Flexible Metal Conduit: Type FMC

350 Liquidtight Flexible Metal Conduit: Type LFMC

352 Rigid Nonmetallic Conduit: Type RNC

353 High Density Polyethylene Conduit: Type HDPE Conduit

354 Nonmetallic Underground Conduit with Conductors: Type NUCC

355 Reinforced Thermosetting Resin Conduit: Type RTRC

356 Liquidtight Flexible Nonmetallic Conduit: Type LFNC

358 Electrical Metallic Tubing: Type EMT

360 Flexible Metallic Tubing: Type FMT

362 Electrical Nonmetallic Tubing: Type ENT

366 Auxiliary Gutters

368 Busways

370 Cablebus

372 Cellular Concrete Floor Raceways

374 Cellular Metal Floor Raceways

376 Metal Wireways

378 Nonmetallic Wireways

380 Multioutlet Assembly

382 Nonmetallic Extensions

384 Strut Type Channel Raceway

386 Surface Metal Raceways

388 Surface Nonmetallic Raceways

390 Underfloor Raceways

392 Cable Trays

394 Concealed Knob and Tube Wiring

396 Messenger Supported Wiring

398 Open Wiring on Insulators

Chapter 4—Equipment for General Use

The scope of Chapter 4 is Information and Rules on Equipment for General Use in Electrical Installations. Articles in this chapter addressing equipment for general use include information on all equipment in an electrical installation. Note that Chapter 3 addressed "Wiring Methods and Materials." Chapter 4 addresses "electrical equipment" necessary for utilization, control, generation, and transformation of electrical energy in an electrical installation. This chapter includes, for example, requirements for equipment cords/cables, switches, receptacles, panelboards, generators, transformers, appliances, motors, and other utilization equipment. It has 21 articles:

400 Flexible Cords and Cables

402 Fixture Wires

404 Switches

DidYouKnow?

A review of the articles in Chapters 1 through 4 reveals that these are the basic building blocks for all electrical installations and are therefore the basis upon which each electrical installation is built.

406 Receptacles, Cord Connectors, and Attachment Plugs (Caps)

408 Switchboards and Panelboards

409 Industrial Control Panels

410 Luminaires (Lighting Fixtures), Lampholders, and Lamps

411 Lighting Systems Operating at 30 Volts or Less

422 Appliances

424 Fixed Electric Space Heating Equipment

426 Fixed Outdoor Electric De-icing and Snow Melting Equipment

427 Fixed Electric Heating Equipment for Pipelines and Vessels

430 Motors, Motor Circuits, and Controllers

440 Air Conditioning and Refrigerating Equipment

445 Generators

450 Transformers and Transformer Vaults (Including Secondary Ties)

455 Phase Converters

460 Capacitors

470 Resistors and Reactors

480 Storage Batteries

490 Equipment Over 600 Volts, Nominal

The requirements of Chapters 1 through 4 will apply for the electrical installation in your home as well as all dwelling units, including single- and multifamily dwelling units.

FIGURE 3–1 Chapters 1 through 4 form the foundation for all basic electrical installations, including dwelling units.

The requirements of Chapters 1 through 4 will apply for the electrical installation in commercial occupancies such as a food market, drugstore, bakery, shopping mall, or office space.

As required in Section 90.3, Chapters 5, 6, and 7 supplement or modify the first four chapters. Unlike the first four, which are general for all electrical installations, Chapters 5, 6, and 7 deal with special requirements. They apply to special occupancies, special equipment, or special conditions.

Chapter 5—Special Occupancies

The scope of this chapter is Modifications and/or Supplemental Information/Rules for Electrical Installations in Special Occupancies. This chapter includes "occupancy-specific" information supplementing or modifying the first four chapters in special occupancies, such as hazardous locations, health care facilities, places of assembly, carnivals, agricultural buildings, mobile homes, RVs, and marinas. This chapter contains 27 articles:

DidYouKnow?

The "occupancy-specific" requirements of Chapter 5 modify or supplement the general rules in Chapters 1 through 4.

500 Hazardous (Classified) Locations, Classes I, II, and III, Divisions 1 and 2

501 Class I Locations

502 Class II Locations

503 Class III Locations

504 Intrinsically Safe Systems

505 Class I, Zone 0, 1, and 2 Locations

506 Zone 20, 21, and 22 Locations for Combustible Dusts, Fibers, and Flyings

510 Hazardous (Classified) Locations, Specific

511 Commercial Garages, Repair and Storage

513 Aircraft Hangars

514 Motor Fuel Dispensing Facilities

515 Bulk Storage Plants

516 Spray Application, Dipping and Coating Processes

517 Health Care Facilities

518 Assembly Occupancies

520 Theatres, Audience Areas of Motion Picture and Television Studios, Performance Areas, and Similar Locations

522 Control Systems for Permanent Amusement Attractions

525 Carnivals, Circuses, Fairs, and Similar Events

530 Motion Picture and Television Studios and Similar Locations

540 Motion Picture Projection Rooms

545 Manufactured Buildings

547 Agricultural Buildings

550 Mobile Homes, Manufactured Homes, and Mobile Home Parks

551 Recreational Vehicles and Recreational Vehicle Parks

552 Park Trailers

553 Floating Buildings

555 Marinas and Boatyards

590 Temporary Installations

The requirements in Chapters 1 through 4 will apply for the electrical installation in special occupancies such as a hospital, a marina, a refinery,

FIGURE 3-2 Chapters 1 through 4 form the foundation for all basic electrical installations, including commercial and institutional occupancies.

DidYouKnow?

It is essential that the user of the *NEC*® be familiar with the special occupancies addressed in Chapter 5.

industry, farms, etc. with the requirements of Chapter 5 supplementing or modifying those rules.

Chapter 6—Special Equipment

The scope of Chapter 6 is Modifications and/or Supplemental Information/ Rules for Electrical Installations containing Special Equipment. This chapter includes "equipment-specific" information supplementing or modifying

FIGURE 3–3 Chapters 1 through 4 form the foundation for all basic electrical installations, including hospitals, marinas, gas stations, and other Special Occupancies covered in Chapter 5.

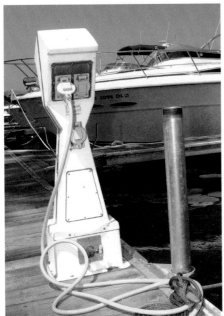

the first four chapters for special equipment, such as electric signs, welders, x-ray equipment, swimming pools, solar photovoltaic systems, fuel cells, and fire pumps. This chapter includes 23 articles:

600 Electric Signs and Outline Lighting

604 Manufactured Wiring Systems

605 Office Furnishings (Consisting of Lighting Accessories and Wired Partitions)

610 Cranes and Hoists

620 Elevators, Dumbwaiters, Escalators, Moving Walks, Wheelchair Lifts, and Stairway Chair Lifts

625 Electric Vehicle Charging System

626 Electrified Truck Parking Space

630 Electric Welders

640 Audio Signal Processing, Amplification, and Reproduction Equipment

645 Information Technology Equipment

647 Sensitive Electronic Equipment

650 Pipe Organs

660 X-Ray Equipment

665 Induction and Dielectric Heating Equipment

668 Electrolytic Cells

669 Electroplating

670 Industrial Machinery

675 Electrically Driven or Controlled Irrigation Machines

680 Swimming Pools, Fountains, and Similar Installations

682 Natural and Artificially Made Bodies of Water

685 Integrated Electrical Systems

690 Solar Photovoltaic Systems

692 Fuel Cell Systems

695 Fire Pumps

The requirements of Chapters 5, 6, and 7 will supplement and/or modify Chapters 1 through 4 to accommodate specific requirements for an electrical installation with special equipment (Chapter 6) such as electric signs, welders, x-ray equipment, swimming pools, solar photovoltaic systems, fuel cells, and fire pumps.

Chapter 7 — Special Conditions

This chapter's scope is Modifications and/or Supplemental Information/ Rules for Electrical Installations containing Special Conditions. This chapter includes "condition-specific" information supplementing or modifying the first four chapters for special conditions, such as emergency systems, legally required standby systems, Class 1 systems, Class 2 systems, Class 3 systems, and fire alarm systems. It has 10 articles:

700 Emergency Systems

701 Legally Required Standby Systems

702 Optional Standby Systems

705 Interconnected Electric Power Production Sources

708 Critical Operations Power System (COPS)

720 Circuits and Equipment Operating at Less Than 50 Volts

725 Class 1, Class 2, and Class 3 Remote Control, Signaling, and Power Limited Circuits

727 Instrumentation Tray Cable: Type ITC

760 Fire Alarm Systems

FIGURE 3–4   Chapters 1 through 4 form the foundation for all basic electrical installations, including Special Equipment listed in Chapter 6.

(continued)

FIGURE 3–4 (continued) Chapters 1 through 4 form the foundation for all basic electrical installations, including Special Equipment listed in Chapter 6.

770 Optical Fiber Cables and Raceways

780 Closed Loop and Programmed Power Distribution

The requirements of Chapters 5, 6, and 7 will supplement and modify Chapters 1 through 4 to accommodate specific requirements for an electrical installation with special conditions (Chapter 7) such as emergency systems, legally required standby systems, Class 1 systems, Class 2 systems, Class 3 systems, and fire alarms.

Chapter 8—Communications Systems

Chapter 8 covers communications systems and is not subject to the requirements of Chapters 1 through 7 except where the requirements are specifically referenced in Chapter 8. This means that all of the articles listed in Chapter 8 stand alone and are not subject to the rules in the rest of the *NEC*® unless a Chapter 8 article specifically references a requirement elsewhere in the *NEC*®. It includes specific information for communications systems, such as communications circuits, radio equipment, television equipment, CATV, and broadband systems. This chapter has four articles:

800 Communications Circuits

810 Radio and Television Equipment

FIGURE 3–5 Chapters 1 through 4 form the foundation for all basic electrical installations, including Special Conditions listed in Chapter 7.

DidYouKnow?

A standby generator may be used to supply an alternate source of power to meet the requirements for Articles: 700 Emergency Systems or, 701 Legally Required Standby Systems or, 702 Optional Standby System.

FIGURE 3-6   Chapter 8 is dedicated to Communications Systems.

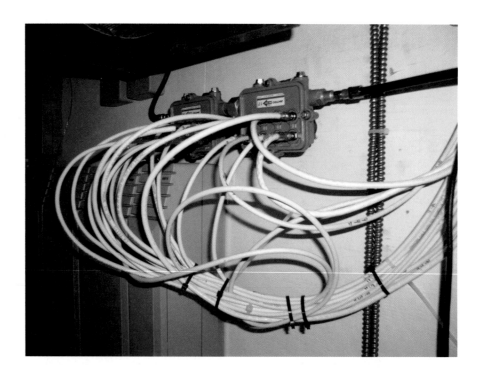

820 Community Antenna Television and Radio Distribution Systems

830 Network-Powered Broadband Communications Systems

Chapter 9—Tables and Annexes A through H

Chapter 9 contains **tables** that are referenced throughout the *NEC*®. **Annexes** are not part of the requirements of the *NEC*® but are included for informational purposes only. The tables and annexes in Chapter 9 are as follows:

Table 1 Percent of Cross Section of Conduit and Tubing for Conductors, *Conduit Fill*

Table 2 Radius of Conduit and Tubing Bends

Table 4 Dimensions and Percent Area of Conduit and Tubing (Areas of Conduit or Tubing for the Combinations of Wires Permitted in Table 1, Chapter 9)

Table 5 Dimensions of Insulated Conductors and Fixture Wires

Table 5A Compact Aluminum Building Wire Nominal Dimensions and Areas

Table 8 Conductor Properties

Table 9 Alternating Current Resistance and Reactance for 600-Volt Cables, 3-Phase, 60 Hz, 75°C (167°F)-Three Single Conductors in Conduit

Table 11A Class 2 & Class 3 Alternating Current Power Source Limitations

Table 11B Class 2 & Class 3 Direct Current Power Source Limitations

Table 12A PLFA Alternating Current Power Source Limitations

Table 12B PLFA Direct Current Power Source Limitations

?

DidYouKnow?

Tables located in Chapter 9 apply only as referenced elsewhere in the *NEC*®. Annexes are informational only.

FIGURE 3-7 Chapter 8 stands alone and is not subject to the requirements of Chapters 1 through 7 unless a specific reference is made in a Chapter 8 article.

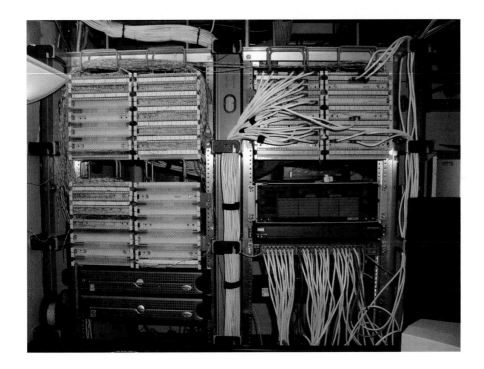

Annex A Product Safety Standards

Annex B Application Information for Ampacity Calculation

Annex C Conduit and Tubing Fill Tables for Conductors and Fixture Wires of the Same Size

Annex D Examples

Annex E Types of Construction

Annex F Availability and Reliability for Critical Operations Power Systems; and Development and Implementation of Functional Performance Tests (FPTs) for Critical Operations Power Systems

Annex G Supervisory Control and Data Acquisition (SCADA)

Annex H Administration and Enforcement

SUMMARY

One of the most basic and crucial steps toward being proficient in the use of the *National Electrical Code®* is an in-depth understanding of how the Code is arranged. In Article 90, the Introduction to the *NEC®*, Section 90.3 explains the Code arrangement in detail as follows:

- The *NEC®* is divided into the Introduction and nine chapters.
- Chapters 1, 2, 3, and 4 apply generally to all electrical installations.
- Chapters 5, 6, and 7 are Special. These apply to Special Occupancies, Special Equipment, and Special Conditions. These Special Chapters supplement and/or modify the general rules in Chapters 1 through 4.

- Chapter 8 stands alone and is not subject to the requirements of Chapters 1 through 7 unless there is a specific reference in a Chapter 8 article.
- Chapter 9 contains tables that are applicable only when referenced.
- Annexes are informational only and are not mandatory.

An in-depth understanding of this arrangement is necessary to properly apply the requirements of the *NEC®*. Understanding the arrangement of the *NEC®* and using the table of contents as a starting point for all inquiries into the *NEC®* is one of the cornerstones of the Codeology method.

REVIEW QUESTIONS

1. The tables located in Chapter 9 of the *National Electrical Code®* apply _____.
   a. at all times
   b. only in Chapters 5, 6, and 7
   c. wherever they are useful
   d. as referenced in the *NEC®*

2. Which chapter of the *National Electrical Code®* stands alone and is not subject to other chapters of the *NEC®* unless a specific reference is made?
   a. Chapter 1
   b. Chapter 5
   c. Chapter 8
   d. Chapter 9

3. Chapters 1 through 4 of the *National Electrical Code®* apply _____ to all electrical installations.
   a. without modification
   b. generally
   c. sparingly
   d. in some cases

4. Annexes in Chapter 9 of the *National Electrical Code®* are _____.
   a. informational only
   b. mandatory requirements
   c. applicable as referenced
   d. used only for special equipment

5. Chapters 5, 6, and 7 of the *National Electrical Code®* will _____ Chapters 1 through 4.
   a. not be associated in any way with
   b. apply generally with
   c. not have an effect on
   d. supplement or modify

6. The arrangement of the *NEC®* is outlined in section _____ of the *NEC®*.
   a. 90.1
   b. 110.3(B)
   c. 210.8
   d. 90.3

7. The requiremens of Chapter 3 apply to Chapter 8 Articles only when _____.

   a. the job is over 200,000 squre feet

   b. a specific reference is made in a Chapter 8 Article

   c. the inspector says so

   d. the requirement sure seems to fit the installation

8. Chapters 1 though 4 will apply generally in _____.

   a. hospitals

   b. single-family homes

   c. marinas

   d. all of the above

9. Chapter 6 deals specifically with special _____.

   a. equipment

   b. people

   c. projects

   d. problems

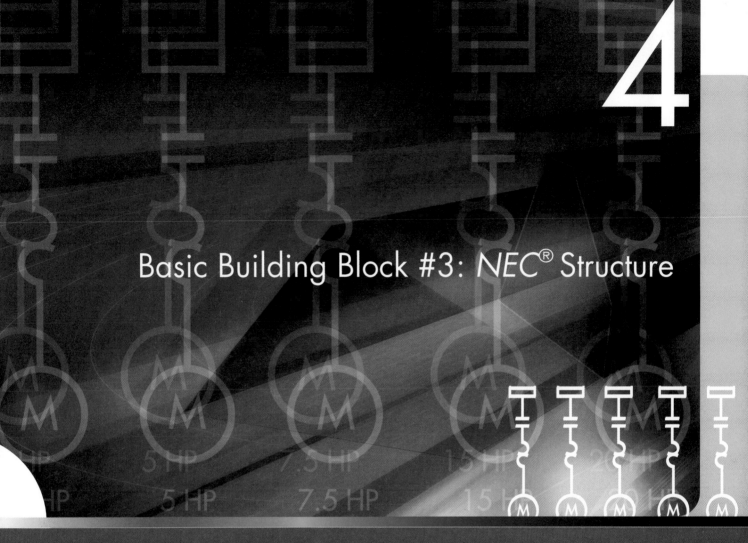

4

Basic Building Block #3: *NEC®* Structure

OUTLINE

OBJECTIVES

After completing this unit, you should be able to:
1. Outline the structure of the *NEC*® from chapter to articles, parts, sections, three levels of subdivisions, exceptions, and list items
2. Recognize the importance of the separation of articles into parts
3. Recognize the application of different levels of subdivisions within a section
4. Identify the mandatory and permissive text within the *NEC*®
5. Recognize that Fine Print Notes are informational only
6. Recognize the location and importance of cross-reference tables
7. Recognize the usefulness of outlines, diagrams, and drawings within the *NEC*®
8. Identify the application of annexes, units of measurement, and extract material

OVERVIEW

This unit will introduce the third of four basic building blocks that are essential for the understanding and application of the Codeology method. Throughout this text, these basic principles will be repeated time and time again until they become part of our natural thought process when working with the *NEC*®.

When one opens the *NEC*® or any other type of installation code, it is absolutely necessary to understand the structure of the code. The structure of the *NEC*® is governed and outlined by the *National Electrical Code*® Style Manual.

NEC® ORGANIZATION

The *National Electrical Code®* is an extremely organized installation document. The rules that govern the structure of the *NEC®* are known as the *National Electrical Code®* **Style Manual.** These rules are available online at www.nfpa.org from the National Fire Protection Agency. The *NEC®* Style Manual is intended to be used as a practical working tool to assist in making the *NEC®* as clear, usable, and unambiguous as possible. In examining the structure of the *NEC®*, the starting point is the table of contents.

The Outline Form of the NEC®: Chapters

As previously stated, the table of contents is broken down into 10 separate pieces. The first is the Introduction to the *NEC®*, Article 90. The *NEC®* is then subdivided into chapters that are major subdivisions covering broad areas divided into articles. Nine chapters follow the Introduction to the *NEC®*. The first eight chapters outline a number of articles under the title and scope of the chapter. Chapter 9 is broken down into tables and annexes. Articles are chapter **subdivisions** covering a specific subject such as branch circuits, grounding, transformers, rigid metal conduit, motors, etc. Each article is given an individual title. Articles are then divided into **sections** and sometimes parts. When an article is sufficiently large, it is sometimes subdivided into parts that correspond to logical groupings of information. Parts are given individual titles and are designated by Roman numerals. Parts are then divided into sections. Sections are sometimes divided into up to three levels of subdivisions to clarify a requirement. Sections and the first two levels of subdivisions are always given a title. Sections and subdivisions may contain lists, **exceptions,** and Fine Print Notes (FPNs) (Figure 4–1).

The Outline Form of the NEC®: Articles

Articles are subdivisions of chapters that cover a specific topic within the scope of the chapter. For example, Chapter 1 is titled "General" and the articles within this chapter must be of a general nature for all electrical installations. The articles contained in Chapter 1, the 100 series, are general in nature and are titled, "Article 100, Definitions" and "Article 110, Requirements for Electrical Installations."

The Outline Form of the NEC®: Parts

When necessary, due to large size or for usability, an article is subdivided into separate parts, each of which is dedicated to a logical separation of requirements within the given article. For example, "Chapter 1—General" contains "Article 110, Requirements for Electrical Installations." This article is then logically separated into five different parts that are individually titled to describe the requirements contained within the part. Parts are always numbered with Roman numerals. They are:

* Part I General
* Part II 600 Volts, Nominal, or Less

FIGURE 4-1   Hierarchy of the *NEC*®.

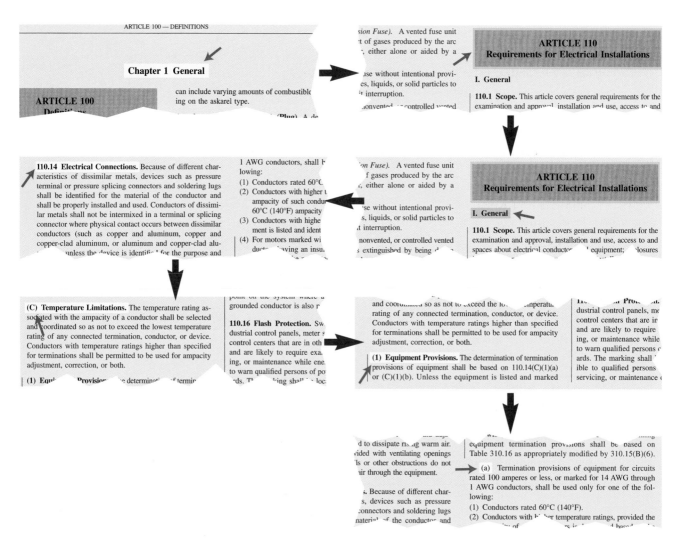

- Part III Over 600 Volts, Nominal
- Part IV Tunnel Installations over 600 Volts, Nominal
- Part V Manholes and Other Electric Enclosures Intended for Personnel Entry, All Voltages

DidYouKnow?

Parts are a logical separation of information within an Article.

The Outline Form of the *NEC*®: Sections

Each part is then subdivided into separate sections, each of which is dedicated to a separate rule under the title of the specific part. These logical separations of requirements into separate sections are individually titled to identify the rule/s and are designed to aid the code user to quickly find the necessary information.

Following is an explanation of how sections are organized and presented:

Sections are always identified by a section number and a title in bold print:

110.14 Electrical Connections

First-level subdivisions are always identified by an upper-case letter in parentheses and a title in bold print:

(C) Temperature Limitations

Second-level subdivisions are identified by a number in parentheses and a title in bold print:

(1) Equipment Provisions

Third-level subdivisions are identified by a lower-case letter in parentheses, may be titled, not in bold print:

(a) Termination provisions of equipment for circuits rated 100 amperes or less, or marked for 14 AWG through 1 AWG conductors, shall be used only for one of the following:

List items are identified by a lower-case letter or number in parentheses followed by the text, without title, not in bold print:

(1) Conductors rated 60°C (140°F)

Exceptions are identified by italicized text, not in bold print, and are numbered where more than one exists. An *exception* will immediately follow the section, subdivision, or list item to which it applies.

Fine Print Notes are identified by the acronym "FPN" and are followed by the text of the Fine Print Note. A FPN will immediately follow the section, subdivision, or list item to which it applies. Remember FPNs are informational only and not an enforceable part of the *NEC*®:

FPN: With respect to 110.14(C)(1) and (2), equipment markings or listing information may additionally restrict the sizing and temperature ratings of connected conductors.

DidYouKnow?

Sections, first-level subdivisions, and second-level subdivisions are always in bold print and are always provided with a title for usability.

Application of Rules

The structure of the *NEC*® is in a progressive ladder-type format. For example, a rule that exists in a third-level subdivision applies only under the second level subdivision. In addition, the second-level subdivision is limited to the rule in the first-level subdivision, which applies only under the section in which it exists, which applies only in the part it is arranged in, which applies only in the article in which it exists, which is limited to the chapter in which it is located.

The key to properly applying the rules of the *NEC*® is to always apply the rule within the part of the article in which it exists. Without an understanding of the outline form of the *NEC*®, a new or inexperienced user may attempt to broadly apply a rule to areas in which it may not apply. Using the Codeology method, the user will always know which part of what article and chapter the section exists. This basic information is crucial to the proper application of all *NEC*® rules. Table 4–1 reviews the application of rules in the *NEC*®:

DidYouKnow?

Sections may be subdivided into three levels to clearly illustrate the intended requirement.

TABLE 4-1 Application of Rules in the *NEC®*

Section	Applies only within the scope of the part of the article in which it is located.
First-Level Subdivision	Applies only within the scope of the section in which it exists.
Second-Level Subdivision	Applies only within the scope of the first-level subdivision in which it exists.
Third-Level Subdivision	Applies only within the scope of the second-level subdivision in which it exists.
List Items	Applies only in the section or subdivision in which they exist.
Exceptions	Applies only to the section, subdivision, or list item under which they exist.
Fine Print Notes	Used for informational purposes only and are designed to aid the user in the application of the rule/s under which they exist.

Exceptions

Exceptions are only used where they are absolutely necessary. *Exceptions* are always italicized in the *NEC®* for quick and easy identification. The *NEC®* Code-Making Panels strive to make the Code as user-friendly as possible. Most *exceptions* have been eliminated and the rules changed to positive text, which is easier to read and apply. *Exceptions* are used where:

- The rule in which the *exception* is applied is modified or supplemented elsewhere in the *NEC®*. This modification or supplement will be qualified to very specific locations, equipment, conditions, or wiring methods or uses. Examples of this type of *exception* are:
  - 210.8(A)(3) *Exception*
  - 314.28 *Exception*
  - 450.3(B) *Exception*
- Existing conditions may require alternate methods or modification of the rule. Examples of this type of *exception* are:
  - 110.26(E) *Exception*
  - 300.20(A) *Exception No.1*
- *Exceptions* are written to allow specific variations from the general rule to explain and clarify the intent and scope of the rule. Examples of this type of exception are:
  - 110.26(F)(1)(a) *Exception*
  - 310.15(B)(2)(a) *Exceptions 1 through 5*

Where an *exception* to a rule exists, the *exception* will immediately follow the main rule to which it applies. An *exception* is used in the *NEC®* only where necessary. When possible, the Technical Committees (Code-Making Panels) will use positive language within a given section instead of an *exception*. When *exceptions* are made to sections that contain list items, the *exception* will clearly indicate the items within the list to which it applies. Where *exceptions* are used, they truly are an *exception* to the rule they follow. For example:

ARTICLE 300 Wiring Methods

Part I. General

300.12 Mechanical Continuity—Raceways and Cables.

Metal or nonmetallic raceways, cable armors, and cable sheaths shall be continuous between cabinets, boxes, fittings, or other enclosures or outlets.

Exception No. 1: Short sections of raceways used to provide support or protection of cable assemblies from physical damage shall not be required to be mechanically continuous.

The rule stated in 300.12 requires that all raceways, cable armors, and cable sheaths are continuous between boxes, fittings, or other enclosures or outlets.

The following *exception* allows short sections of raceways only (not cable armors and cable sheaths) to provide support and/or protection of cable assemblies. For example, this *exception* would permit a short section of a raceway, such as electrical metallic tubing (EMT) or rigid metal conduit (RMC) to protect and/or support a cable assembly for a short distance where physical damage could occur to the cable assembly. Note that while the *NEC*® does not state a minimum length, the term "short sections" infers a piece of a raceway.

ARTICLE 250 Grounding and Bonding

Part III. Grounding Electrode System and Grounding Electrode Conductor

250.68 Grounding Electrode Conductor and Bonding Jumper Connection to Grounding Electrodes.

(A) Accessibility. All mechanical elements used to terminate a grounding electrode conductor or bonding jumper to a grounding electrode shall be accessible.

Exception No.1: An encased or buried connection to a concrete-encased, driven, or buried grounding electrode shall not be required to be accessible.

Exception No. 2: Exothermic or irreversible compression connections used at terminations, together with the mechanical means used to attach such terminations to fireproofed structural metal whether or not the mechanical means is reversible, shall not be required to be accessible.

The rule stated in 250.68 (A) requires that all connections of grounding electrode conductors to grounding electrodes be accessible. There are two exceptions. *Exception No. 1* allows for connections encased in concrete or buried being inaccessible. *Exception No. 2* allows for exothermic or irreversible connections to structural steel to be covered in fireproofing materials.

DidYouKnow?

Where an exception is used in the NEC, it will immediately follow the rule to which it applies.

Fine Print Notes

Fine Print Notes are explanatory material and are *not an enforceable part of the NEC*® as required in Section 90.5(C). There are different types of FPNs. These FPNs will include but are not limited to the following types:

* Informational
* Reference (referencing other sections or areas of this code or other codes)
* Design

- Suggestions
- Examples

FPNs are used when

- FPN provides basic information to aid the user of the *NEC®*. Examples of this type of FPN include:
  - 90.5(C) FPN
  - Article 100 Definition of "Listed" FPN
  - 110.11 FPN No. 2
- FPN provides example/s of where or how the rule/s would apply. Examples of this type of FPN include:
  - 250.20 FPN
  - 250.94(3) FPN No. 1

- FPN provides reference to other sections within the *NEC®* to further explain the requirement and to aid the user in proper application. Examples of this type of FPN include:
  - 90.7 FPN No. 1, 2 & 3
  - 250.20(D) FPN No. 2
  - 314.15 FPN No. 1 & 2
- FPN provides reference to other codes or standards to further explain the requirement, inform of building code requirements, and to aid the user in proper application. Examples of this type of FPN include:
  - 110.16 FPN No. 1 & 2
  - 210.52 FPN
  - 300.21 FPN
- FPN provides a suggestion for adequate performance and/or proper design. Examples of this type of FPN include:
  - 210.4(A) FPN
  - 210.19(A)(1) FPN No. 4
  - 215.2(A)(3) FPN No. 2
  - 220.61(C) FPN No. 2
  - 240.85 FPN

The following is an example of a FPN:

250.20 Alternating-Current Systems to Be Grounded.

Alternating-current systems *shall* be grounded as provided for in 250.20(A), (B), (C), (D), or (E). Other systems *shall be permitted* to be grounded. If such systems are grounded, they shall comply with the applicable provisions of this article.

FPN: An example of a system permitted to be grounded is a corner-grounded delta transformer connection. See 250.26(4) for conductor to be grounded.

This rule is followed by an informational FPN that gives an example of a system that is permitted to be grounded. The FPN continues in the second sentence with a reference within the *NEC®* to Section 250.26, List Item (4). This is an *informational FPN* in the form of an example and a *reference FPN* within the *NEC®* to aid the user of this code.

Material for FPNs is included in the *NEC®* only where the Code-Making Panel believes that the information is necessary for proper application of the rule/s. In most cases the text of a FPN is indispensable information for the user and the proper application of the *NEC®*. Always read the FPNs.

Tables

Tables, which are located in Chapter 9, are only applicable when referenced in other chapters of the *NEC®*. For example, Table 1 is used for "Conduit Fill" and is applicable only where referenced in a section of the *NEC®*. Table 1 in Chapter 9 will be permitted for use to calculate raceway fill for raceway articles such as, "Article 358 Electrical Metallic Tubing: Type EMT." Article 358 is subdivided into three parts. Part II is titled "Installation" and includes Section 358.22 that references Table 1 in Chapter 9. This specific reference allows the use of the table:

358.22 Number of Conductors.

The number of conductors shall not exceed that permitted by the percentage fill specified in <u>Table 1, Chapter 9.</u>

Cables shall be permitted to be installed where such use is not prohibited by the respective cable articles. The number of cables shall not exceed the allowable percentage fill specified in <u>Table 1, Chapter 9.</u>

DidYouKnow?

Tables located in Chapter 9 of the *NEC®* are applicable only when referenced elsewhere in the code.

Annexes

Annexes were known as Appendixes in the *NEC®* prior to the 2002 edition. As stated in 90.3, annexes are for informational purposes only. The material contained in an annex is provided to aid the user in the understanding and application of the requirements in the *NEC®*. For example, "Annex A Product Safety Standards" contains information to aid the Code user by providing a list of product safety standards used for product listing where that listing is required by the *NEC®*. Annex A is not a part of the requirements of this NFPA document but is included for informational purposes only.

For example, in "Article 348 Flexible Metal Conduit: Type FMC," "Part I General" contains Section 348.6 Listing Requirements that requires all FMC and associated fittings to be listed. Annex A provides the "Product Standard Number" for Flexible Metal Conduit, which is "UL 1."

DidYouKnow?

Annexes provide the *NEC®* user with additional information, such as a cross-reference to applicable Product Standards in Annex A.

MANDATORY RULES, PERMISSIVE RULES, AND EXPLANATORY MATERIAL

The *NEC®* follows specific rules to clearly illustrate whether requirements are mandatory or permissive. These rules also identify text that is explanatory in nature and is provided to aid the user of the code.

The *NEC®* provides the method used to determine the applicability of all text in "Article 90 Introduction." This article is not subdivided into parts. Section 90.5 clearly defines the application of all rules and material in the *NEC®* as follows:

90.5 Mandatory Rules, Permissive Rules, and Explanatory Material.

(A) **Mandatory Rules.** Mandatory rules of this Code are those that identify actions that are specifically required or prohibited and are characterized by the use of the terms shall or shall not.

DidYouKnow?

Alternate methods in the *NEC*® are prefaced or followed by the phrase "shall be permitted."

DidYouKnow?

The term "shall" or "shall not" in the *NEC*® is used for mandatory requirements.

DidYouKnow?

In order to fully understand the structure of the *NEC*®, the Code user must be able to identify mandatory, permissive, and explanatory text.

(B) **Permissive Rules.** Permissive rules of this Code are those that identify actions that are allowed but not required, are normally used to describe options or alternative methods, and are characterized by the use of the terms *shall be permitted* or *shall not be required*.

(C) **Explanatory Material.** Explanatory material such as references to other standards, references to related sections of this Code, or information related to a Code rule, is included in this Code in the form of fine print notes (FPNs). Fine print notes are informational only and are not enforceable as requirements of this Code.

Brackets containing section references to another NFPA document are for informational purposes only and are provided as a guide to indicate the source of the extracted text. These bracketed references immediately follow the extracted text.

FPN: The format and language used in this Code follows guidelines established by NFPA and published in the *NEC*® Style Manual. Copies of this manual can be obtained from NFPA.

Mandatory Language

As stated in 90.5, a rule is considered mandatory by the use of the terms "shall" or "shall not." Where the term "shall" or "shall not" is used in the *NEC*®, the rule is mandatory unless an exception exists or the rule exists in Chapters 1 through 4 and is supplemented or modified in Chapters 5 through 7. Following is an example of mandatory text:

230.6 Conductors Considered Outside the Building.

Conductors ***shall*** be considered outside of a building or other structure under any of the following conditions:

(1) Where installed under not less than 50 mm (2 in.) of concrete beneath a building or other structure

(2) Where installed within a building or other structure in a raceway that is encased in concrete or brick not less than 50 mm (2 in.) thick

(3) Where installed in any vault that meets the construction requirements of Article 450, Part III

(4) Where installed in conduit and under not less than 450 mm (18 in.) of earth beneath a building or other structure

The above rule in 230.6 exists in "Article 230 Services" in "Part I General." This rule requires that service conductors (conductors meeting the definition of "service conductors" in Article 100), which meet any of the conditions in list items 1 through 4, be considered as being outside of the building.

Permissive Text

As stated in 90.5, a rule is considered permissive by the use of the terms "shall be permitted" or "shall not be required." Where the terms "shall be permitted" or "shall not be required" are used in the *NEC*®, the rule is permissive unless an exception exists or the rule exists in Chapters 1 through 4 and is supplemented or modified in Chapters 5 through 7. See the following example of permissive text:

250.20 Alternating-Current Systems to Be Grounded.

Alternating-current systems ***shall*** be grounded as provided for in 250.20(A), (B), (C), (D), or (E). Other systems ***shall be permitted*** to be grounded. If such systems are grounded, they shall comply with the applicable provisions of this article.

FPN: An example of a system permitted to be grounded is a corner-grounded delta transformer connection. See 250.26(4) for conductor to be grounded.

The above rule in 250.20 exists in "Article 250 Grounding" in "Part II System Grounding." This rule requires that:

1. Alternating-current systems *shall* be grounded as provided for in 250.20(A), (B), (C), or (D), and

2. Other systems *shall be permitted* to be grounded. If such systems are grounded, they shall comply with the applicable provisions of this article.

The second requirement of this section specifically permits systems other than those illustrated in 250.20(A), (B), (C), (D), or (E) to be grounded by the use of the term "shall be permitted." This is in essence a mandated permission. The alternative would be for the *NEC®* to use the term "may" to illustrate permissiveness. However, the term "may" is considered unenforceable and is not used. The use of the term "may" would mean that the inspector "may" permit other systems to be grounded or he/she "may not." The *NEC®* Style Manual lists terms to be avoided such as "may" that are vague and/or unenforceable in an effort to make the code as clear and usable as possible.

Table 4–2 summarizes the careful use of language in the *NEC®*.

TABLE 4-2   Text Language

Type of Text	Listed Term/s
Mandatory text	"Shall"
	"Shall not"
Permissive text	"Shall be permitted"
	"Shall not be required"
Explanatory text	Fine Print Notes

Units of Measurement

The **units of measurement** in the *NEC®* uses both the SI system (metric units) and inch-pound units. The SI system is always shown first and inch-pound second in parentheses. See the following example:

334.30 Securing and Supporting.

Nonmetallic-sheathed cable shall be supported and secured by staples, cable ties, straps, hangers, or similar fittings designed and installed so as not to damage the cable at intervals not exceeding **1.4 m (4 1/2 ft.)** and within **300 mm (12 in.)** of every outlet box, junction box, cabinet, or fitting. Flat cables shall not be stapled on edge.

Sections of cable protected from physical damage by raceway shall not be required to be secured within the raceway.

DidYouKnow?

The SI system (metric units) is always shown first in the *NEC®* with the inch-pound units shown second in parentheses because this Code is used internationally.

The rule in 334.30 exists in "Article 334 Nonmetallic-Sheathed Cable: Types NM, NMC, and NMS" in "Part II Installation." This rule requires that Type NM cable be secured and supported. Note that the SI system is always listed first and the inch-pound units second and always in parentheses. Note also that section 90.9(D) specifically permits the use of either the SI system or the inch-pound system.

Extract Material

DidYouKnow?

The *NEC*® includes requirements from other NFPA Codes and Standards in the form of *extract material*.

The *National Electrical Code*®, also known as NFPA-70, is one of many documents published by the National Fire Protection Association. Where another NFPA document has primary jurisdiction over material to be included in the *NEC*®, the material is extracted into the *NEC*®. An example of another NFPA document that would have primary jurisdiction over material addressed by the *NEC*® is "NFPA-20, Standard for the Installation of Stationary Pumps for Fire Protection." When this occurs, the document in which the extract material exists is identified at the beginning of the article. A FPN will immediately follow the title of the article to inform the Code user of the presence of extract material. Rules within the article that are extracted will be followed with the title of the referenced NFPA document and sections in brackets. This reference to sections of another NFPA document is for informational purposes only as stated in 90.5(C). An example of **extract material** in the *NEC*® exists in Article 695:

ARTICLE 695 Fire Pumps

FPN: Rules that are followed by a reference in brackets contain text that has been extracted from NFPA 20-2007, Standard for the Installation of Stationary Pumps for Fire Protection. Only editorial changes were made to the extracted text to make it consistent with this Code.

The FPN following the title of Article 695 informs the user that it contains extract material from "NFPA 20-2007, Standard for the Installation of Stationary Pumps for Fire Protection." The FPN further explains that where this material is located, it is identified with a reference in brackets at the end of the rule. This extracted material can only be changed editorially to fit the style of the *NEC*®. The following section from Article 695 contains extracted material as follows:

DidYouKnow?

Cross-reference tables provide the Code user with areas in the *NEC*® which may modify or supplement a given requirement.

ARTICLE 695 Fire Pumps

695.10 Listed Equipment.

Diesel engine fire pump controllers, electric fire pump controllers, electric motors, fire pump power transfer switches, foam pump controllers, and limited service controllers shall be listed for fire pump service. [NFPA 20: 9.5.1.1, 10.1.2.1, 12.1.3.1]

The brackets located at the end of 695.10 identify the sections and title of the document from which the extract material originated.

Cross-Reference Tables

There are six **cross-reference tables** included in the *NEC*®. These tables are provided where there are multiple modifications or supplemental requirements related to the scope of an article located elsewhere in the *NEC*®. Table 4–3 shows where the cross-reference tables are located and Figure 4–2 shows an example.

TABLE 4–3 Cross-Reference Tables

Type of Text	Listed Term/s
ARTICLE 210 Branch Circuits Table 210.2 Specific-Purpose Branch Circuits	Provides cross-references to aid the Code user in 32 other locations.
ARTICLE 220 Branch Circuit Feeder and Service Calculations Table 220.3 Additional Load Calculation References	Provides cross-references to aid the Code user in 29 other locations.
ARTICLE 225 Outside Branch Circuits and Feeders 225.2 Other Articles	Provides cross-references to aid the Code user in 25 other articles.
ARTICLE 240 Overcurrent Protection Table 240.3 Other Articles	Provides cross-references to aid the Code user in 35 other articles.
ARTICLE 250 Grounding and Bonding Table 250.3 Additional Grounding Requirements	Provides cross-references to aid the Code user in over 80 other locations.
ARTICLE 430 Motors, Motor Circuits, and Controllers Table 430.5 Other Articles	Provides cross-references to aid the Code user in 25 other locations.

Outlines, Diagrams, and Drawings

The *NEC®* does not contain pictures to aid the Code user in application of installation requirements. Section 90.1(C) clearly states that the *NEC®* is not intended as a design manual or as an instruction manual for untrained persons. However, the *NEC®* does include **outlines, diagrams,** and **drawings** that describe the application of an article or section or provide basic information in ladder-type diagrams to aid the Code user. These informational outlines are provided as follows:

90.3 Code Arrangement

The ladder diagram provided with this section is designed to clearly outline the arrangement and application of rules contained in the *NEC®*.

210.52 Dwelling Unit Receptable Outlets

Figure 210.52 (C) (1) Determination of Area Behind a Range, or Counter-Mounted Cooking Unit or Sink

This drawing is provided to aid the user of this code in applying the receptacle outlet requirement of 210.52(C)(1) near a sink, counter-mounted cooking, unit or range.

220.1 Scope, Figure 220.1 Branch Circuit, Feeder and Service Calculation Methods

This ladder diagram is included to aid the code user in understanding the permitted calculation methods of Article 220.

230.1 Scope, Figure 230.1 Services

The diagram provided in Figure 230.1 is a useful outline of how to apply *NEC®* rules for services from the service point to the premises wiring. (See Figure 4–3 for an example of a figure in the *NEC®*.)

FIGURE 4–2

Example of cross-reference table.

Equipment	Article
Air-conditioning and refrigerating equipment	440
Appliances	422
Audio signal processing, amplification, and reproduction equipment	640
Branch circuits	210
Busways	368
Capacitors	460
Class 1, Class 2, and Class 3 remote-control, signaling, and power-limited circuits	725
Closed-loop and programmed power distribution	780
Cranes and hoists	610
Electric signs and outline lighting	600
Electric welders	630
Electrolytic cells	668
Elevators, dumbwaiters, escalators, moving walks, wheelchair lifts, and stairway chair lifts	620
Emergency systems	700
Fire alarm systems	760
Fire pumps	695
Fixed electric heating equipment for pipelines and vessels	427
Fixed electric space-heating equipment	424
Fixed outdoor electric deicing and snow-melting equipment	426
Generators	445
Health care facilities	517
Induction and dielectric heating equipment	665
Industrial machinery	670
Luminaires (lighting fixtures), lampholders, and lamps	410
Motion picture and television studios and similar locations	530
Motors, motor circuits, and controllers	430
Phase converters	455
Pipe organs	650
Places of assembly	518
Receptacles	406
Services	230
Solar photovoltaic systems	690
Switchboards and panelboards	408
Theaters, audience areas of motion picture and television studios, and similar locations	520
Transformers and transformer vaults	450
X-ray equipment	660

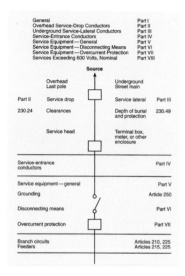

DidYouKnow?

The drawing in Figure 680.8, which provides a visual illustration of clearances from a pool, together with Table 680.8, is an excellent example of how figures and drawings in the *NEC*® aid the Code user.

250.1 Scope

Figure 250.1 Grounding and Bonding

This ladder diagram details the organization of Article 250 Grounding and Bonding.

430.1 Scope, Figure 430.1 Article 430 contents.

The single-line diagram provided in Figure 430.1 is a useful outline of how to apply *NEC*® rules for motors.

514.3 Classification of Locations, Figure 514.3 Classified Areas Adjacent to Dispensers as Detailed in Table 514.3(B)(1) [NFPA 30A: Figure 8.3.1]

This drawing is included to aid the Code user in applying the rules of Article 514 in classified locations adjacent to dispensers.

515.3 Class 1 Locations, Figure 515.3 Marine Terminal Handling Flammable Liquids [NFPA 30: Figure 7.7.16]

This drawing is included to aid the Code user in applying the rules of Article 515 in classified locations on marine terminals.

516.3(B)(1), (2), (4), and (5)

Figures 516.3(B)(1), (2), (4), and (5)

These drawings are included to aid the Code user in applying the rules of Article 516 for spray areas, rooms, and booths.

517.30, Figures No. 1 & 2

These single-line diagrams provide a useful tool for the Code user to understand and apply the rules of Article 517 for Essential Electrical Systems for Hospitals.

517.41, Figures No. 1 & 2

These single-line diagrams provide a useful tool for the Code user to understand and apply the rules of Article 517 for Nursing Homes and Limited Health Care Facilities.

550.10(C) Attachment Plug Cap, Figure 550.10(C)

551.46(C) Attachment Plugs, Figure 551.46(C)

552.44(C) Attachment Plugs, Figure 552.44(C)

These similiar drawings depict required configurations for attachment plug caps and grounding-type receptacles for the following articles: 550 Mobile Homes, Manufactured Homes, and Mobile Home Parks; 551 Recreational Vehicles and Recreational Vehicle Parks; and 552 Park Trailers.

620.2 Control System, Figure 620.2 Control System

This ladder diagram illustrates the elevator control system.

620.13 Feeder and Branch Circuit Conductors, Figure 620.13 Single Line Diagram

The single-line diagram provided in Figure 620.13 is a useful outline of how to apply *NEC*® rules for elevators.

680.8 Overhead Conductor Clearances, Figure 680.8 Clearances From Pool Structures

The drawing in Figure 680.8 together with Table 680.8 provide the Code user with a visual explanation of overhead clearances.

690.1 Scope, Figures 690.1(A) & (B)

The drawings and diagrams of Figure 690.1(A) and Figure 690.1(B) help the Code user identify solar photovoltaic system components as well as common system configurations.

725.121 Power Sources for Class 2 and Class 3 Circuits, Figure 725.121 Class 2 and Class 3 Circuits

Figure 725.121 illustrates the relationships between Class 2 or Class 3 power sources, their supply, and the Class 2 or Class 3 circuits.

725.61 Applications of Listed Class 2, Class 3, and PLTC Cables, Figure 725.61 Cable Substitution Hierarchy

760.154 Applications of Listed PLFA Cables, Figure 760.154 Cable Substitution Hierarchy

770.154 Applications of Listed Optical Fiber Cables and Raceways, Figure 770.154(E) Cable Substitution Hierarchy

800.154 Applications of Listed Communications Wires and Cables and Communications Raceways, Figure 800.154(E) Cable Substitution Hierarchy

820.154 Applications of Listed CATV Cables and CATV Raceways, Figure 820.154(E) Cable Substitution Hierarchy

These drawings help describe cable substitution hierarchy.

SUMMARY

One of the most basic and crucial steps toward becoming proficient in the use of the National Electrical Code® is an in-depth understanding of the structure of the text within the *NEC*®. The structure of the *NEC*® is governed by the *NEC*® Style Manual. This manual provides a detailed outline form followed consistently throughout the *NEC*®. An understanding of this outline form is essential to the proper application of *NEC*® requirements. The structure of the *NEC*® is summarized as follows:

- The *NEC*® is subdivided into 10 major subdivisions, the Introduction and nine chapters.
- Chapters, which are major subdivisions of the *NEC*®, are charged with broad scopes. The scope of each chapter is subdivided logically into separate articles to address each chapter scope.
- Articles are major subdivisions of chapters that cover a specific topic within the scope of the chapter. When an article is of a large size or for usability, it is subdivided into separate parts, each of which is dedicated to a logical separation of requirements within the given article.
- Parts are major subdivisions of articles that logically separate information for ease of use and proper application. Parts are then broken down into separate sections individually titled to address the scope of the individual part.
- Sections may be logically subdivided into three levels. Sections may also contain list items, exceptions, and FPNs.
- The structure of the *NEC*® is in a progressive ladder-type format, which when applied is as follows:
    - a rule that exists in a *third-level subdivision* applies only under:
    - the rule in the *second-level subdivision*, which is limited to:
    - the rule in the *first-level subdivision*, which applies only under:
    - the *section* in which it exists, which applies only in:
    - the *part* it is arranged in, which applies only in the:
    - *article* in which it exists, which is limited to the:
    - *chapter* in which it is located.
- List items are used in sections, subdivisions, or exceptions where necessary.
- Exceptions are used only when absolutely necessary and are always italicized.
- FPNs are informational only and are not mandatory.
- Tables in Chapter 9 are applicable only where referenced elsewhere in the *NEC*®.
- Annexes are informational only and are not mandatory.
- Mandatory text in the *NEC*® consists of the use of "shall" and "shall not."
- Permissive text in the *NEC*® consists of the use of "shall be permitted" and "shall not be required."
- Explanatory text exists in the form of FPNs.
- Cross-references, outlines, drawings, and diagrams are included to aid the user in the proper application of the *NEC*®.

An in-depth understanding of the structure or outline form of the *NEC*® is necessary to properly apply the requirements of the *NEC*®. Understanding this structure or outline form of the *NEC*® is one of the cornerstones of the Codeology method.

REVIEW QUESTIONS

1. Where there are multiple modifications or supplemental requirements related to the scope of an article located elsewhere in the *NEC®*, an article will contain a _____ table to aid the Code user.

   a. calculations
   b. contents
   c. conduit fill
   d. cross-reference

2. An example of permissive language in the *National Electrical Code®* is which one of the following?

   a. "may be permitted"
   b. "if the installer desires"
   c. "shall be permitted"
   d. "may if cost is an issue"

3. Exceptions are used in the *National Electrical Code®* _____.

   a. Only when necessary
   b. To confuse the Code user
   c. To justify shortcuts
   d. To allow alternate methods

4. FPNs are used in the *National Electrical Code®* to aid the Code user and are _____.

   a. Informational only
   b. Explanatory material
   c. Designed to aid the Code user
   d. All of the above

5. When an article is subdivided into logical separations, these subdivisions are called _____ .

   a. sections
   b. parts
   c. subdivisions
   d. annexes

6. Parts are subdivided into logical separations called _____.

   a. Sections
   b. Parts
   c. Subdivisions
   d. Annexes

7. An example of mandatory language in the *National Electrical Code®* is which one of the following?

   a. "shall not"
   b. "you really should not"
   c. "may not"
   d. "must not"

8. Types of FPNs designed to aid the user of the *National Electrical Code®* include the following:

   a. Informational
   b. Reference
   c. Design suggestions and/or examples
   d. All of the above

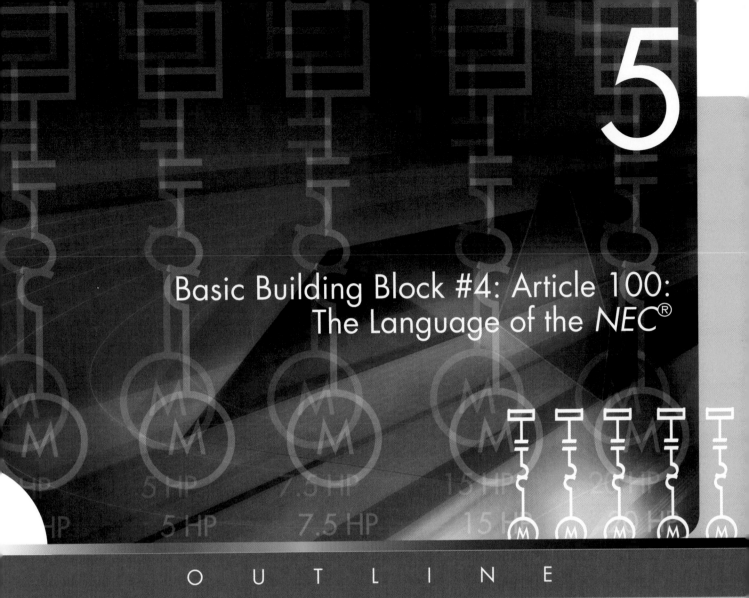

Basic Building Block #4: Article 100:
The Language of the *NEC*®

OBJECTIVES

After completing this unit, you should be able to:
1. Recognize that the *NEC®* defines terms so that the user may understand the language used in the standard and properly apply requirements
2. Recognize that the *NEC®* defines terms in Article 100 only when the definition is essential to proper application
3. Recognize that the *NEC®* does not define common terms or technical terms that are commonly understood
4. Recognize that definitions are located in Article 100 only when the term defined is used in two or more articles
5. Recognize that definitions do not contain requirements
6. Recognize that terms needing to be defined, which exist in a single article, are located in the .2 or second section of the article

OVERVIEW

This unit will introduce the fourth of four basic building blocks that are essential for the understanding and application of the Codeology method. Throughout this text, these basic principles will be repeated time and time again until they become part of our natural thought process when working with the *NEC®*.

When you open the *NEC®* or any other type of installation code, it is absolutely necessary to understand the language of the Code. The language of the *NEC®* is outlined by definitions in Article 100 and, in some cases, definitions within an individual article.

NEC® LANGUAGE: ARTICLE 100

Article 100 contains only those definitions essential to the proper application of this Code. It is not intended to include commonly defined general terms or commonly defined technical terms from related codes and standards. In general, only those terms that are used in two or more articles are defined in Article 100. Other definitions are included in the article in which they are used but may be referenced in Article 100.

Part I of this article contains definitions intended to apply wherever the terms are used throughout this Code. Part II contains definitions applicable only to the parts of articles specifically covering installations and equipment operating at over 600 volts, nominal.

NEC® Language: Definitions

New as well as experienced users of the *National Electrical Code®* must learn the **definitions** for terms defined within this installation document. In order to "walk the walk" one must be able to "talk the talk." The *NEC®* can seem to be written in a different language. **Terms** that are defined must follow the *NEC®* Style Manual. The *NEC®* does not attempt to define commonly used general terms or commonly used technical terms from related codes and standards. Only terms that are essential to the proper application of the *NEC®* are defined. In general, only those terms that are used in two or more articles are defined in Article 100. Other definitions are included in the article in which they are used but may be referenced in Article 100. Definitions are presented as follows:

- All definitions are in alphabetical order
- Definitions shall not contain the term being defined
- Definitions shall not contain requirements or recommendations
- If a term that is defined appears in two or more articles, it is listed in Article 100
- If a term that is defined appears in a single article, it shall be in the second section of that article

All users of the *NEC®* must be familiar with the definitions in order to properly interpret and apply each requirement. The *NEC®* definition for what seems to be a "common term" may differ slightly from a standard dictionary explanation. This can have a tremendous impact on an electrical installation, for example:

Question A metal-sheathed cable assembly such as type AC cable is installed above a lay-in type acoustical ceiling. The cable is installed in accordance with the *NEC®* and the ceiling tiles are replaced. The AC cable can no longer be seen as it is hidden behind the lay-in tiles of the drop ceiling. Is the AC cable considered to be "exposed"?

The answer is "Yes." The AC cable above the lay-in type acoustical ceiling is installed in accordance with the *NEC®* as "exposed." We would normally consider that the term *exposed* meant that we could see the item being discussed. However, the *NEC®* defines the term *exposed* for two different scenarios. The first is *Exposed* (as applied to live parts) and the sec-

DidYouKnow?

In order to understand the language of the *NEC*®, the Code user must be familiar with all definitions in Article 100 as well as those located in individual articles.

ond is *Exposed* (as applied to wiring methods). This question is about type AC cable, which is a wiring method. The following definition would apply:

> **Article 100   Exposed (as applied to wiring methods).** On or attached to the surface or behind panels designed to allow access.

Note that this definition would apply to the type AC cable installed above a lay-in type acoustical ceiling because it is installed behind panels [lay-in ceiling tile] designed to allow access. The term *exposed* is used over 300 times in the *NEC*®. All users of the *NEC*® must be familiar with the most common terms used throughout the document.

Many articles define terms for use only within the same article. The *NEC*® Style Manual requires that terms defined for use in a specific article be in the second section of the article. This is done in the second section of an article so that the user of the *NEC*® will see early on the terms defined for the proper application of the following requirements. The first section is always the scope of an article and the second is for definitions where they are used. For example, Article 517 deals with "Health Care Facilities." The second section of the article is "517.2 Definitions." There are 39 terms defined in this section. Remember that a term that needs to be defined in the *NEC*® is located in Article 100 if the term is used in two or more articles. Article 100 contains over 150 definitions.

The following Article 100 definitions are examples of common terms used in the *NEC*®, which must be clearly understood by the Code user for proper application of the *NEC*®.

- **Accessible, Readily (Readily Accessible).** Capable of being reached quickly for operation, renewal, or inspections without requiring those to whom ready access is requisite to climb over or remove obstacles or to resort to portable ladders, and so forth.

  The term *readily accessible* is used 70 times in the *NEC*®. Other definitions related to accessibility defined in Article 100 are Accessible (as applied to equipment), Accessible (as applied to wiring methods), and Concealed (Figure 5–1).

- **Ampacity.** The current, in amperes, that a conductor can carry continuously under the conditions of use without exceeding its temperature rating.

  The term *ampacity* is used over 400 times in the *NEC*®. This term is derived from the combination of the terms *ampere* and *capacity* (Figure 5–2).

- **Approved.** Acceptable to the authority having jurisdiction.

  The term *approved* is used over 300 times in the *NEC*®. It is essential that the Code user know that where this term is used, compliance is determined by the "authority having jurisdiction" (AHJ), not a third-party listing organization such as UL. The term "authority having jurisdiction" is also defined in Article 100.

- **Branch Circuit.** The circuit conductors between the final overcurrent device protecting the circuit and the outlet(s).

  For proper application of the *NEC*®, the Code user must be able to determine the proper *NEC*® term for all current-carrying conductors. Types of current-carrying conductors are branch circuit, feeder, service, and tap conductors.

FIGURE 5–1

Person operating equipment that is readily accessible.

FIGURE 5-2   Conductor *ampacity* is determined by the *NEC®*.

FIGURE 5-3A

A snap switch. Courtesy of Pass and Seymour/ Legrand.

- **Building.** A structure that stands alone or that is cut off from adjoining structures by fire walls with all openings therein protected by approved fire doors.

  The *NEC®* contains specific requirements and limitations for buildings.

  It is essential to understand that a row of ten strip stores, for example, may be recognized in the *NEC®* as 10 separate buildings, provided they are separated by fire walls.

- **Device.** A unit of an electrical system that carries or controls electric energy as its principal function.

  The term "device" is used in the *NEC®* over 500 times. Types of devices would include but not be limited to receptacles and switches (Figure 5–3a and 5–3b).

- **Exposed (as applied to wiring methods).** On or attached to the surface or behind panels designed to allow access.

  The term "*exposed*" as used in the *NEC®* is very different from the standard use of this term. Other related definitions in Article 100 include "Exposed (as applied to live parts)" and "Concealed."

- **Feeder.** All circuit conductors between the service equipment, the source of a separately derived system, or other power supply source and the final branch-circuit overcurrent device.

  For proper application of the *NEC®*, the Code User must be able to determine the proper *NEC®* term for all current-carrying conductors. Types of current-carrying conductors are branch circuit, feeder, service, and tap conductors.

- **In Sight From (Within Sight From, Within Sight).** Where this Code specifies that one equipment shall be "in sight from," "within sight from," or "within sight of," and so forth, of another equipment, the specified equipment is to be visible and not more than 15 m (50 ft) distant from the other.

  When the *NEC®* requires that equipment, such as a motor, and an associated disconnecting means be "within sight of" each other, the requirement is that the equipment be visible and not more than 50 feet apart.

FIGURE 5-3B

A receptacle. Courtesy of Pass and Seymour/ Legrand.

- **Outlet.** A point on the wiring system at which current is taken to supply utilization equipment.

  The term *outlet* would include a receptacle, a ceiling mounted box for a lighting fixture, and the point at which equipment is hard wired (Figure 5–4).

FIGURE 5–4 Lighting outlet in a ceiling.

- **Overcurrent.** Any current in excess of the rated current of equipment or the ampacity of a conductor. It may result from overload, short circuit, or ground fault.

  An overcurrent includes the following:

  Overload: above the normal full-load rating; for example, 22-amps flowing on a 20-amp branch circuit is an overload.

  Short Circuit: when the current does not flow through its normal path (it takes a short cut) but continues on circuit conductors, it is said to be a short circuit.

  Ground Fault: a type of short circuit in which current flows outside of the circuit conductors and returns to the source through grounded equipment or an equipment grounding conductor.

- **Qualified Person.** One who has skills and knowledge related to the construction and operation of the electrical equipment and installations and has received safety training to recognize and avoid the hazards involved.

  The term *qualified person* is used over 100 times in the *NEC*®. It is extremely important for all Code users to understand that the qualified person has specific skills and knowledge as well as safety training.

- **Service Conductors.** The conductors from the service point to the service disconnecting means.

For proper application of the *NEC*®, the Code user must be able to determine the proper *NEC*® term for all current-carrying conductors. Types of current-carrying conductors are branch circuit, feeder, service, and tap conductors.

SUMMARY

One of the most basic and crucial steps toward being proficient in the use of the *National Electrical Code*® is an in-depth understanding of the language spoken in the *NEC*®. The *NEC*® does not define all commonly used terms. Terms that are essential to the proper application of the *NEC*® are defined. When a term is defined and it is used in two or more articles, the definition is placed in Article 100. When a term is defined but used in a single article, it is placed in the second section of the article. Definitions do not contain requirements and are listed in alphabetical order. Reading and applying the *NEC*® can be as difficult as understanding a foreign language without an in-depth understanding of definitions within the *NEC*®. Understanding this language, the terms defined in the *NEC*®, is one of the cornerstones of the Codeology method.

REVIEW QUESTIONS

1. The *National Electrical Code*® defines only terms that are essential for _____ of the *NEC*®.
   a. spelling terms
   b. changing
   c. printing
   d. proper application

2. The lack of understanding of a term defined in the *NEC*® will result in the following:
   a. Misapplication of the *NEC*®
   b. Confusion
   c. Violations
   d. All of the above

3. Definitions are placed in Article 100 only when the term being defined is used in _____ or more articles.
   a. four
   b. two
   c. six
   d. three

4. Where a defined term exists in only one article, the term is defined in the _____ section of the article.
   a. second
   b. largest
   c. most appropriate
   d. first

5. There are _____ definitions which contain the term "service."
   a. seven
   b. eight
   c. nine
   d. ten

6. Definitions are not permitted to contain _____.
   a. the term being defined
   b. a requirement
   c. a & b
   d. none of the above

7. Definitions are always placed in _____ order.
   a. chronological
   b. numerical
   c. alphabetical
   d. reverse

8. Definitions in Article 100 apply _____ in the NEC.
   a. globally
   b. sparingly
   c. intermittenly
   d. optionally

6

Codeology Fundamentals

OUTLINE

OBJECTIVES

After completing this unit, you should be able to:
1. Recognize the importance of customizing your codebook with notes, highlighting articles and parts, underlining, and code tabs
2. Begin to identify clues and key words to locate the proper chapter, article, and part within the *NEC*®
3. Identify the four basic building blocks of Codeology
4. Recognize the Codeology outline of the *NEC*®
5. Implement the fundamentals of Codeology

OVERVIEW

This unit will introduce the fundamentals of the Codeology method. Previous units have reviewed in detail the four building blocks that form the foundation of the Codeology method. With a basic understanding of the Code arrangement, the table of contents, the structure and language of the *NEC*®, the systematic approach to the Codeology method is now illustrated. This method is described in detail with an outline of the Codeology method that is applied to the table of contents and the fundamental steps to apply this system. This unit will also describe how to customize your codebook, enhancing the Codeology method, through the use of notes, highlighting, and tabs.

MARKING UP YOUR CODEBOOK

A new copy of the *NEC®* is unmarked in any way. Users of the codebook should put their name on the side or inside cover to personalize and identify it as their own. The *NEC®* is not printed in color; implementing the Codeology method is enhanced by marking up your codebook. The following suggestions will make your copy of the *NEC®* easier to use and will enhance your ability to quickly and accurately find needed information using the Codeology method.

Mark Up Your Table of Contents

The table of contents is the starting point for all inquiries in the *NEC®*. While each major subdivision of the table of contents is titled, it is extremely helpful to clarify the scope with notes (Figure 6–1). Print the following notes, defining chapter scope, in the table of contents.

The Article 90	*The* Introduction *and ground rules for the NEC®*
Chapter 1	General *Information and Rules for Electrical Installations*
Chapter 2 *Plan*	*Information and Rules on* Wiring and Protection *of Electrical Installations*
Chapter 3 *Build*	*Information and Rules on* Wiring Methods and Materials *for use in Electrical Installations*
Chapter 4 *use*	*Information and Rules on* Equipment for General Use *for use in Electrical Installations*
Chapter 5	*Modifications and/or Supplemental Information/Rules for Electrical Installations in* Special Occupancies
Chapter 6	*Modifications and/or Supplemental Information/Rules for Electrical Installations containing* Special Equipment
Chapter 7	*Modifications and/or Supplemental Information/Rules for Electrical Installations containing* Special Conditions
Chapter 8	Communications Systems *only*

DidYouKnow?

Customizing the Table of Contents aids the Code user by illustrating the broad scope of each Chapter of the *NEC®*.

Highlighting

Highlighters are an extremely useful tool. The number one rule with highlighting is to be careful to highlight only where absolutely necessary. Some users of the *NEC®* use highlighters liberally, sometimes highlighting an entire page. This can be counterproductive, as when this material is revisited, the user is moved to read the entire page that is highlighted instead of applying the Codeology method. Many different color highlighters are available and the

FIGURE 6-1 The table of contents marked up.

CONTENTS

Contents

NATIONAL ELECTRICAL CODE   2005 Edition

user of this Code should use those colors to their advantage. When material is highlighted for future reference, it should be limited and yellow in color.

Highlighting Articles and Parts

The most productive use of highlighters is to identify each article and part in order to have them stand out as we move through the *NEC*®. The following exercise is mandatory for all Codeology students and requires a green and an orange highlighter. The color green will be used to highlight all articles. The color orange will be used to highlight all parts. These colors should not be used for any other highlighting in your codebook. This will allow the Codeology user to instantly recognize the beginning of a new part or a new article. This is extremely important, because many Code users will start in the right part of the right article, but will move into the next part of the article to find an answer. While green and orange are used for marking only articles and parts, yellow should be used, only when necessary, for highlighting other code text.

The following exercise requires a green and an orange highlighter.

- Open your *NEC*® to the table of contents.
- Begin with Article 90 and determine the page number from the table of contents. Highlight the article number and title in green.
  - ARTICLE 90 Introduction
- Move on to Article 100 and determine the page number from the table of contents. Highlight the article number and title in green and the parts in orange as follows:
  - ARTICLE 100 Definitions
  - I. General
  - II. Over 600 Volts, Nominal
- Move on to Article 110 and determine the page number from the table of contents. Highlight the article number and title in green and the parts in orange as follows:
  - ARTICLE 110 Requirements for Electrical Installations
    - I. General
    - II. 600 Volts, Nominal, or Less
    - III. Over 600 Volts, Nominal
    - IV. Tunnel Installations over 600 Volts, Nominal
    - V. Manholes and Other Electric Enclosures Intended for Personnel Entry, All Voltages

Continue through the table of contents, finding the page number for each article (highlight in green) and each part of an article (highlight in orange).

Marking each article in green and each part in orange will enable you to quickly identify both articles and parts of articles. When applying the Codeology method, it is imperative to know at all times what part of what article a section is in to properly apply the requirement (Figure 6–2).

Tabs

Code tabs are an extremely effective tool to quickly find articles and sections of the *NEC*® that are used most frequently. When applying code tabs be sure to take your time and apply them properly. Read and follow the instructions provided with your code tabs.

Underlining

Another effective tool when using the *NEC*® is to underline sections, subdivisions, list items, or any material to bring it to your attention in the future. Together with highlighting, underlining of important text will be extremely useful when revisiting a specific area of the *NEC*® (Figure 6–3).

DidYouKnow?

The proper application of the Codeology method requires that the code user locate Articles and Parts very quickly. By using highlighters to identify Articles in green and Parts in orange, the Code user will be able to quickly and confidently get to the right section.

FIGURE 6–2   *NEC*® Article 410, Part I and Part II highlighted.

410.1   ARTICLE 410 — LUMINAIRES (LIGHTING FIXTURES), LAMPHOLDERS, AND LAMPS

---

(2) Supply voltage, phase, frequency, and full-load current.

(3) Short-circuit current rating of the industrial control panel based on one of the following:

    a. Short-circuit current rating of a listed and labeled assembly

    b. Short-circuit current rating established utilizing an approved method

    FPN: UL 508A-2001, Supplement SB, is an example of an approved method.

(4) If the industrial control panel is intended as service equipment, it shall be marked to identify it as being suitable for use as service equipment.

(5) Electrical wiring diagram or the number of the index to the electrical drawings showing the electrical wiring diagram.

(6) An enclosure type number shall be marked on the industrial control panel enclosure.

**ARTICLE 410
Luminaires (Lighting Fixtures),
Lampholders, and Lamps**

I. General

410.1 Scope. This article covers luminaires (lighting fixtures), lampholders, pendants, incandescent filament lamps, arc lamps, electric-discharge lamps, decorative lighting products, lighting accessories for temporary seasonal and holiday use, portable flexible lighting products, and the wiring and equipment forming part of such products and lighting installations.

410.2 Application of Other Articles. Equipment for use in hazardous (classified) locations shall conform to Articles 500 through 517. Lighting systems operating at 30 volts or less shall conform to Article 411. Arc lamps used in theaters shall comply with 520.61, and arc lamps used in projection machines shall comply with 540.20. Arc lamps used on constant-current systems shall comply with the general requirements of Article 490.

410.3 Live Parts. Luminaires (fixtures), lampholders, and lamps shall have no live parts normally exposed to contact. Exposed accessible terminals in lampholders and switches shall not be installed in metal luminaire (fixture) canopies or in open bases of portable table or floor lamps.

Exception: Cleat-type lampholders located at least 2.5 m (8 ft) above the floor shall be permitted to have exposed terminals.

II. Luminaire (Fixture) Locations

410.4 Luminaires (Fixtures) in Specific Locations.

(A) Wet and Damp Locations. Luminaires (fixtures) installed in wet or damp locations shall be installed so that water cannot enter or accumulate in wiring compartments, lampholders, or other electrical parts. All luminaires (fixtures) installed in wet locations shall be marked, "Suitable for Wet Locations." All luminaires (fixtures) installed in damp locations shall be marked, "Suitable for Wet Locations" or "Suitable for Damp Locations."

(B) Corrosive Locations. Luminaires (fixtures) installed in corrosive locations shall be of a type suitable for such locations.

(C) In Ducts or Hoods. Luminaires (fixtures) shall be permitted to be installed in commercial cooking hoods where all of the following conditions are met:

(1) The luminaire (fixture) shall be identified for use within commercial cooking hoods and installed such that the temperature limits of the materials used are not exceeded.

(2) The luminaire (fixture) shall be constructed so that all exhaust vapors, grease, oil, or cooking vapors are excluded from the lamp and wiring compartment. Diffusers shall be resistant to thermal shock.

(3) Parts of the luminaire (fixture) exposed within the hood shall be corrosion resistant or protected against corrosion, and the surface shall be smooth so as not to collect deposits and to facilitate cleaning.

(4) Wiring methods and materials supplying the luminaire(s) [fixture(s)] shall not be exposed within the cooking hood.

    FPN: See 110.11 for conductors and equipment exposed to deteriorating agents.

(D) Bathtub and Shower Areas. No parts of cord-connected luminaires (fixtures), chain-, cable-, or cord-suspended-luminaires (fixtures), lighting track, pendants, or ceiling-suspended (paddle) fans shall be located within a zone measured 900 mm (3 ft) horizontally and 2.5 m (8 ft) vertically from the top of the bathtub rim or shower stall threshold. This zone is all encompassing and includes the zone directly over the tub or shower stall. Luminaires (lighting fixtures) located in this zone shall be listed for damp locations, or listed for wet locations where subject to shower spray.

(E) Luminaires (Fixtures) in Indoor Sports, Mixed-Use, and All-Purpose Facilities. Luminaires (fixtures) subject to physical damage, using a mercury vapor or metal halide lamp, installed in playing and spectator seating areas of indoor sports, mixed-use, or all-purpose facilities shall be

Notes for Specific Section Application

As your Code studies progress to article- or topic-specific courses, take the time to write brief, neat notes in *pencil* for future reference. Use the blank

FIGURE 6-3   Section 230.70 with a single sentence underlined in red pencil.

ARTICLE 230 — SERVICES   **230.72**

Exception: For jacketed multiconductor service cable without splice.

(F) Drip Loops. Drip loops shall be formed on individual conductors. To prevent the entrance of moisture, service-entrance conductors shall be connected to the service-drop conductors either (1) below the level of the service head or (2) below the level of the termination of the service-entrance cable sheath.

(G) Arranged That Water Will Not Enter Service Raceway or Equipment. Service-drop conductors and service-entrance conductors shall be arranged so that water will not enter service raceway or equipment.

230.56 Service Conductor with the Higher Voltage to Ground. On a 4-wire, delta-connected service where the midpoint of one phase winding is grounded, the service conductor having the higher phase voltage to ground shall be durably and permanently marked by an outer finish that is orange in color, or by other effective means, at each termination or junction point.

V. Service Equipment — General

230.62 Service Equipment — Enclosed or Guarded. Energized parts of service equipment shall be enclosed as specified in 230.62(A) or guarded as specified in 230.62(B).

(A) Enclosed. Energized parts shall be enclosed so that they will not be exposed to accidental contact or shall be guarded as in 230.62(B).

(B) Guarded. Energized parts that are not enclosed shall be installed on a switchboard, panelboard, or control board and guarded in accordance with 110.18 and 110.27. Where energized parts are guarded as provided in 110.27(A)(1) and (A)(2), a means for locking or sealing doors providing access to energized parts shall be provided.

230.66 Marking. Service equipment rated at 600 volts or less shall be marked to identify it as being suitable for use as service equipment. Individual meter socket enclosures shall not be considered service equipment.

VI. Service Equipment — Disconnecting Means

230.70 General. Means shall be provided to disconnect all conductors in a building or other structure from the service-entrance conductors.

(A) Location. The service disconnecting means shall be installed in accordance with 230.70(A)(1), (A)(2), and (A)(3).

(1) Readily Accessible Location. The service disconnecting means shall be installed at a readily accessible location either outside of a building or structure or inside <u>nearest the point of entrance of the service conductors.</u>

(2) Bathrooms. Service disconnecting means shall not be installed in bathrooms.

(3) Remote Control. Where a remote control device(s) is used to actuate the service disconnecting means, the service disconnecting means shall be located in accordance with 230.70(A)(1).

(B) Marking. Each service disconnect shall be permanently marked to identify it as a service disconnect.

(C) Suitable for Use. Each service disconnecting means shall be suitable for the prevailing conditions. Service equipment installed in hazardous (classified) locations shall comply with the requirements of Articles 500 through 517.

230.71 Maximum Number of Disconnects.

(A) General. The service disconnecting means for each service permitted by 230.2, or for each set of service-entrance conductors permitted by 230.40, Exception Nos. 1, 3, 4, or 5, shall consist of not more than six switches or sets of circuit breakers, or a combination of not more than six switches and sets of circuit breakers, mounted in a single enclosure, in a group of separate enclosures, or in or on a switchboard. There shall be not more than six sets of disconnects per service grouped in any one location. For the purpose of this section, disconnecting means used solely for power monitoring equipment, transient voltage surge suppressors, or the control circuit of the ground-fault protection system or power-operable service disconnecting means, installed as part of the listed equipment, shall not be considered a service disconnecting means.

(B) Single-Pole Units. Two or three single-pole switches or breakers, capable of individual operation, shall be permitted on multiwire circuits, one pole for each ungrounded conductor, as one multipole disconnect, provided they are equipped with handle ties or a master handle to disconnect all conductors of the service with no more than six operations of the hand.

> FPN: See 408.36(A) for service equipment in panelboards, and see 430.95 for service equipment in motor control centers.

230.72 Grouping of Disconnects.

(A) General. The two to six disconnects as permitted in 230.71 shall be grouped. Each disconnect shall be marked to indicate the load served.

Exception: One of the two to six service disconnecting means permitted in 230.71, where used only for a water pump also intended to provide fire protection, shall be permitted to be located remote from the other disconnecting means.

DidYouKnow?

Underlining in the color red is extremely useful for the Code user to highlight requirements for future reference.

pages in the back of the *NEC*® for additional notes. This is an ideal place for the Code user to make notes for possible proposals to change the next edition of the *NEC*® (Figure 6–4).

FIGURE 6-4   Notes to Section 430.6.

Controller. For the purpose of this article, a controller is any switch or device that is normally used to start and stop a motor by making and breaking the motor circuit current.

Motor Control Circuit. The circuit of a control apparatus or system that carries the electric signals directing the performance of the controller but does not carry the main power current.

System Isolation Equipment. A redundantly monitored, remotely operated contactor-isolating system, packaged to provide the disconnection/isolation function, capable of verifiable operation from multiple remote locations by means of lockout switches, each having the capability of being padlocked in the "off" (open) position.

430.4 Part-Winding Motors. A part-winding start induction or synchronous motor is one that is arranged for starting by first energizing part of its primary (armature) winding and, subsequently, energizing the remainder of this winding in one or more steps. A standard part-winding start induction motor is arranged so that one-half of its primary winding can be energized initially, and, subsequently, the remaining half can be energized, both halves then carrying equal current. A hermetic refrigerant compressor motor shall not be considered a standard part-winding start induction motor.

Where separate overload devices are used with a standard part-winding start induction motor, each half of the motor winding shall be individually protected in accordance with 430.32 and 430.37 with a trip current one-half that specified.

Each motor-winding connection shall have branch-circuit short-circuit and ground-fault protection rated at not more than one-half that specified by 430.52.

Exception: A short-circuit and ground-fault protective device shall be permitted for both windings if the device will allow the motor to start. Where time-delay (dual-element) fuses are used, they shall be permitted to have a rating not exceeding 150 percent of the motor full-load current.

430.5 Other Articles. Motors and controllers shall also comply with the applicable provisions of Table 430.5.

Always start at 430.6

430.6 Ampacity and Motor Rating Determination. The size of conductors supplying equipment covered by Article 430 shall be selected from the allowable ampacity tables in accordance with 310.15(B) or shall be calculated in accordance with 310.15(C). Where flexible cord is used, the size of the conductor shall be selected in accordance with 400.5. The required ampacity and motor ratings shall be determined as specified in 430.6(A), (B), and (C).

(A) General Motor Applications. For general motor applications, current ratings shall be determined based on (A)(1) and (A)(2).

Table 430.5 Other Articles

Equipment/Occupancy	Article	Section
Air-conditioning and refrigerating equipment	440	
Capacitors		460.8, 460.9
Commercial garages; aircraft hangars; motor fuel dispensing facilities; bulk storage plants; spray application, dipping, and coating processes; and inhalation anesthetizing locations	511, 513, 514, 515, 516, and 517 Part IV	
Cranes and hoists	610	
Electrically driven or controlled irrigation machines	675	
Elevators, dumbwaiters, escalators, moving walks, wheelchair lifts, and stairway chair lifts	620	
Fire pumps	695	
Hazardous (classified) locations	500–503 and 505	
Industrial machinery	670	
Motion picture projectors		540.11 and 540.20
Motion picture and television studios and similar locations	530	
Resistors and reactors	470	
Theaters, audience areas of motion picture and television studios, and similar locations		520.48
Transformers and transformer vaults	450	

(1) Table Values. Other than for motors built for low speeds (less than 1200 RPM) or high torques, and for multispeed motors, the values given in Table 430.247, Table 430.248, Table 430.249, and Table 430.250 shall be used to determine the ampacity of conductors or ampere ratings of switches, branch-circuit short-circuit and ground-fault protection, instead of the actual current rating marked on the motor nameplate. Where a motor is marked in amperes, but not horsepower, the horsepower rating shall be assumed to be that corresponding to the value given in Table 430.247, Table 430.248, Table 430.249, and Table 430.250, interpolated if necessary. Motors built for low speeds (less than 1200 RPM) or high torques may have higher full-load currents, and multispeed motors will have full-load current varying with speed, in which case the nameplate current ratings shall be used.

Exception No. 1: Multispeed motors shall be in accordance with 430.22(A) and 430.52.

CLUES AND KEY WORDS

Clues will come in many forms and each step of the Codeology method exposes clues for individual chapters, articles, and parts. The use of clues or key words is essential to qualify your question or need to get into the right part of the right article in the right chapter. Clues are found throughout the following:

- Questions
- Chapter Titles
- Article Titles
- Titles of Parts
- Titles of Sections and Subdivisions

  Clues or key words will include, but are not limited to, the following:

Basic/General	Chapter 1
Plan	Chapter 2
Build	Chapter 3
Use	Chapter 4
Specials	Chapters 5, 6, and 7
Communications	Chapter 8
Occupancy types	
Indoor/Outdoor	
Computed/Calculated	
Never/Always	
Outdoors	
Temporary	
Over 600 volts	

  Clues and key words will be examined in subsequent chapters of the Codeology text, focusing on each chapter in the table of contents.

THE FOUR BASIC BUILDING BLOCKS OF THE CODEOLOGY METHOD

The previous units have built the foundation for this method to be introduced and applied. The four basic building blocks provide the basis for our Codeology method. They can be summarized as outlined below:

  Building Block #1
- **The Table of Contents**
  - The outline of the *NEC*®
  - Ten major subdivisions of the *NEC*®
    - The Introduction, Article 90
    - Chapters 1 through 9
  - Chapters subdivided into articles within the scope of each chapter

  Building Block #2
- **90.3 Code Arrangement**
  - Chapters 1 through 4 apply generally to all installations
  - Chapters 5, 6, and 7 supplement or modify Chapters 1 through 4
  - Chapter 8 stands alone unless a specific reference exists
  - Chapter 9 tables apply as referenced; annexes are informational only

Did You Know?

The four basic building blocks of the Codeology method are an understanding of:
- *NEC*® Table of Contents
- 90.3 Code Arrangement
- The structure of the *NEC*®
- Language, Definitions in the *NEC*®

Building Block #3
- **The structure of the *NEC*®**
  - Ten major subdivisions of the *NEC*®
    - The Introduction, Article 90
    - Chapters 1 through 9
  - Each chapter covers a broad area
  - Each chapter is subdivided into articles to address chapter scope
  - Articles are subdivided logically into parts
  - Parts are subdivided into sections
  - Sections can contain three levels of subdivisions
  - Sections and subdivisions can contain exceptions
  - Sections and subdivisions can contain list items
  - Sections and subdivisions can contain FPNs
  - Mandatory language: shall or shall not
  - Permissive language: shall be permitted or shall not be required
  - Informational material contained in FPNs and annexes

Building Block #4
- **Definitions, the language of the *NEC*®**
  - Terms used in more than one article are defined in Article 100
  - Terms used in a single article are defined in the second section of the article

GETTING IN THE RIGHT CHAPTER, ARTICLE, AND PART

The Codeology method by design allows a user to determine exactly where in the *NEC*® to begin to look for the section which addresses his/her needs. This method begins by applying the Codeology outline and getting in the right chapter. Through the use of key words and clues, the user identifies the correct article and part in which to begin his/her inquiry into the *NEC*®. These are the fundamentals of Codeology.

Introduction to the Fundamentals of Codeology

The Codeology method is designed to teach the student a systematic, disciplined approach to quickly find information by understanding and applying the outline form of the *NEC*®. The use of generic terms to aid the Codeology user in finding the right chapter in the table of contents is essential. The generic terms include Plan, Build, and Use. This portion of the Codeology text will briefly introduce the basic steps of Codeology, which will then be explained in detail in following units.

The following is a basic Codeology outline defining the basic steps for finding needed information in the *NEC*® quickly and accurately. This outline will use several general terms to help qualify the question or need in order to quickly identify the correct chapter. Clues and key words help identify the correct chapter.

The general terms used to steer the Codeology user in the right direction are Basic/General, Plan, Build, Use, Specials, and Communications. The main focus will be on Chapters 1 through 8. The Introduction in Article 90

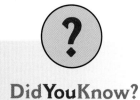

DidYouKnow?

Chapter 1 contains basic or general requirements applicable to all electrical installations.

lays the ground rules for the use of the *NEC®*; the tables in Chapter 9 will be used only where referenced elsewhere in the codebook. The use of these terms in the Codeology method is described below:

General

When a question or need for information in the *NEC®* is basic or general in nature to all electrical installations, think **General** and start in "Chapter 1— General" (Figure 6–5). Chapter 1 contains two articles that address "Definitions" and "Requirements for Electrical Installations." Key words and clues that should steer the Codeology user to think "General, Chapter 1" include:

- A definition question
- Examination, installation, and use of equipment (listed, labeled)
- Mounting and cooling of equipment
- Electrical connections
- Flash protection
- Identification of disconnecting means
- Workspace clearance

FIGURE 6–5   The general requirements of Chapter 1 include provisions for adequate working space for persons and dedicated space for electrical equipment.

DidYouKnow?

Before we can build an electrical installation, we must have a plan. Chapter 2 is identified in the Codeology Method as the "Plan" chapter.

Plan

When a question or need for information in the *NEC®* deals with planning stages general in nature to all electrical installations, think ***Plan*** and start in

"Chapter 2—Wiring and Protection" (Figure 6–6). The title of Chapter 2 includes two terms: *wiring* and *protection.*

FIGURE 6-6   Chapter 2 of the *NEC*® is called the "Plan" chapter. All electrical installations must be properly planned.

The term *wiring,* as used in Chapter 2, does not mean different types of cable assemblies or raceways. Wiring in Chapter 2 addresses the *NEC*® terms for the types of current-carrying conductors common to all electrical installations. These wiring terms include branch circuits, feeders, services, taps, and transformer secondary conductors. Chapter 2 provides the basic requirements for all of these conductors, including calculations to properly size them.

The term *protection,* as used in Chapter 2, includes the basic protection requirements for all electrical installations. This protection includes overcurrent protection, grounding, bonding, surge arrestors, and surge-protective devices. Key words and clues that should steer the Codeology user to think "Plan, Chapter 2" include:

- Branch circuit
- Feeder
- Service
- Calculation or computed load
- Overcurrent protection
- Grounding
- Surge arrestors
- Surge-protective devices (SPDs)

Build

When a question or need for information in the *NEC*® deals with building an electrical installation, think ***Build*** and start in Chapter 3. All inquiries on methods and materials to get from the source of energy to the load (all physical wiring of an installation) will require wiring methods and/or materials that are in Chapter 3—Wiring Methods and Materials (Figure 6–7). The title of Chapter 3 includes two terms: *wiring methods* and *wiring materials.* These two terms cover all means and methods of electrical distribution. This chapter does not cover panelboards, disconnects, transformers, or utilization equipment. However, this chapter does cover every physical means of wiring branch circuits, feeders, services, taps, and transformer secondary conductors.

?

DidYouKnow?

Building an electrical installation requires methods and materials to get current from point A to Point B. Chapter 3 is identified in the Codeology Method as the "Build" chapter.

FIGURE 6–7   Chapter 3, the "Build" chapter, includes wiring methods and wiring materials to get electrical current from point A to point B in all electrical installations.

The "Wiring Methods" portion of Chapter 3 includes general information for all wiring methods and materials as well as specific articles for all conductors, cable assemblies, raceways, busways, cablebus, etc. In addition, the wiring materials portion of Chapter 3 includes general information for all wiring materials as well as specific articles for all boxes (outlet, device, pull, junction), conduit bodies, cabinets, cutout boxes, auxiliary gutters, and wireways. Key words and clues that should steer the Codeology user to think "Build, Chapter 3" include:

- General questions on wiring methods or materials
- General questions for conductors, uses permitted, ampacity, etc.
- Installation, uses permitted, or construction of any cable assembly
- Installation, uses permitted, or construction of any raceway

DidYouKnow?

Equipment which uses or facilitates the use of electrical energy is located in Chapter 4, which is identified in the Codeology Method as the "Use" chapter.

- Installation, uses permitted, or construction of any other distribution method
- Requirements for boxes of all types
- Requirements for cabinets, meter sockets, wireways, etc.

Use

When a question or need for information in the *NEC*® deals in general with electrical equipment that uses, controls, or transforms electrical energy, think *Use* and start in Chapter 4. All inquiries for information on electrical equipment that controls, transforms, utilizes, or aids in the utilization of electrical energy is in Chapter 4—Equipment for General Use (Figure 6–8).

FIGURE 6–8   Chapter 4 of the *NEC*® is called the "Use" chapter. Utilization equipment, including lighting fixtures and motors, are covered in this chapter along with articles dedicated to the control and installation of utilization equipment.

Chapter 4 includes equipment that uses electrical energy to perform a task, such as lighting fixtures, appliances, heating, de-icing, and snow melting equipment, motors, and air conditioners. In addition, this chapter includes equipment that controls or facilitates the use of electrical energy, such as switches, switchboards, panelboards, cords, cord caps and receptacles. Chapter 4 also covers electrical equipment that generates or tranforms electrical energy, such as generators, transformers, batteries, phase converters, capacitors, resistors, and reactors. Key words and clues that should steer the Codeology user to think "Use, Chapter 4" include:

- Cords or cables
- Lighting fixtures and fixture wires
- Receptacles
- Switchboards and panelboards
- Appliances
- Heating equipment
- Air conditioning equipment
- Motors
- Batteries
- Generators
- Transformers
- Phase converters
- Capacitors

The Specials

When a question or need for information in the *NEC®* deals with special occupancies, equipment, or conditions, think ***Special*** and start in Chapters 5, 6, or 7 (Figure 6–9 a, b, and c). Chapter 1 through 4 are the foundation or backbone for all electrical installations. When an installation contains a "Special Occupancy, Equipment, or Condition," Chapters 5, 6, or 7 will supplement or modify the first four chapters to address the "Special" situation. Key words and clues that should steer the Codeology user to think "Special Occupancies, Chapter 5" include:

- Hazardous, classified locations
- Commercial garages, motor fuel dispensers
- Spray booths and applications
- Hospitals and all health care facilities
- Places of assembly
- Theaters
- Carnivals
- Temporary power
- Manufactured buildings, Motor homes, RVs
- Agricultural buildings, farms
- Marinas, floating buildings

Key words and clues that should steer the Codeology user to think "Special Equipment, Chapter 6" include:

- Electric signs
- Cranes
- Elevators
- Electric welders
- X-ray equipment
- Swimming pools
- Solar and fuel cell systems
- Fire pumps

Did You Know?

The Codeology Method identifies Chapters 5, 6, and 7 as "Special."
- 5 Special Occupancies
- 6 Special Equipment
- 7 Special Conditions

Did You Know?

The "Special" Chapters 5, 6, and 7 will modify and/or supplement the general requirements of Chapters 1 though 4.

FIGURE 6–9 Chapters 5, 6, and 7 of the *NEC*® are the "Special" chapters. Chapter 5 covers Special Occupancies, including hazardous locations. Chapter 6 covers Special Equipment, including pools and spas. Chapter 7 covers Special Conditions, including emergency systems.

(a)

(b)

Continued

FIGURE 6-9   Continued.

(c)

Key words and clues that should steer the Codeology user to think "Special Conditions, Chapter 7" include:

- Emergency systems
- Legally required and optional standby systems
- Class 1, 2, and 3 systems
- Fire alarm systems
- Fiber optics

Communications

When a question or need for information in the *NEC®* deals with communications systems, think ***Communication*** and start in Chapter 8 (Figure 6–10). Chapter 8 is a loner. The rest of the *NEC®* does not apply to this chapter unless there is a specific reference within Chapter 8. Key words and clues that should steer the Codeology user to think "Communications, Chapter 8" include:

    Communications circuits
- Radio and TV
- CATV systems
- Network powered broadband systems

    Table 6–1 illustrates the Codeology Outline of the *NEC®*.

DidYouKnow?

Chapter 8 is identified in the Codeology Method as the "Communications" chapter.

FIGURE 6-10 Chapter 8 of the *NEC*® covers communications systems.

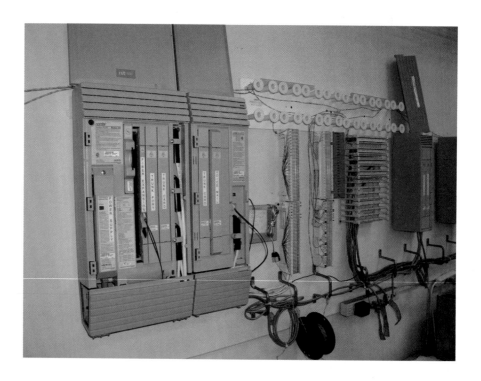

TABLE 6-1 The Codeology Outline

The Ground Rules		
Introduction	Article 90	Introduction/Directions
The Basic Installation		
Chapter 1	100-Series	General
Chapter 2	200-Series	Plan
Chapter 3	300-Series	Build
Chapter 4	400-Series	Use
The Specials		
Chapter 5	500-Series	Occupancies
Chapter 6	600-Series	Equipment
Chapter 7	700-Series	Conditions
Communications: The Loner		
Chapter 8	800-Series	Communications
Tables		
Chapter 9	Tables and Annexes	

FUNDAMENTAL STEPS USING CODEOLOGY

When the need arises to find information in the *NEC*®, you can take several steps to find the answers you are looking for. The following are situations or circumstances when a need for information in the *NEC*® may arise.

- Job site problem/question
- Design concerns
- Inspection questions
- NEC proficiency exam

    The proper steps to get your problems/questions answered are:

Step #1

- Qualify your question or need
- Go to the table of contents
- Look for clues or key words to get into the right chapter
- Get in the right chapter

Step #2

- Further qualify your question or need
- Look for clues or key words
- Use the table of contents and get into the right article within the right chapter

Step #3

- Further qualify your question or need
- Look for clues or key words
- Use the table of contents and get into the right part of the right article

Step #4

- Open the *NEC*® to the part of the article that meets your question or need

Step #5

- Read only the section titles to find the correct section

Step #6

- Read section and all titles of first-level subdivisions
- Read all of the section and pertinent subdivisions including exceptions and FPNs
- Apply the rule or answer your question

SUMMARY

Learning, understanding, and applying the Codeology method begins with four basic building blocks:

- Building block #1 is the table of contents; the Code user must be familiar with and understand the 10 major subdivisions of the *NEC*®.

- Building block #2 is the arrangement of the *NEC*® as detailed in 90.3; the Code user must understand how the individual chapters of the *NEC*® apply.

- Building block #3 is the structure or outline form of the *NEC*®. Without an understanding of the hierarchy of requirements and information in the *NEC*®, proper application would be impossible.

- Building block #4 is the language of the *NEC*®. An in-depth understanding of terms defined is essential for proper application of *NEC*® requirements.

The Codeology outline builds upon the arrangement of the *NEC*® and provides a user-friendly method to help categorize or qualify a question or need through key words and clues. The Codeology outline is as follows:

- The Ground Rules

  Introduction   Article 90   Introduction/
                                          Directions

- The Basic Installation

Chapter 1	100-series	General
Chapter 2	200-series	Plan
Chapter 3	300-series	Build
Chapter 4	400-series	Use

- The Specials

Chapter 5	500-series	Occupancies
Chapter 6	600-series	Equipment
Chapter 7	700-series	Conditions

- Communications, the Loner

  | Chapter 8 | 800-series | Communications |

- Tables

  | Chapter 9 | Tables and Annexes |

Understanding the Codeology outline along with a solid foundation in the four basic building blocks, as described earlier, allows the Code user to begin to apply the Codeology method. Additional fundamentals include the recognition of key words and clues, marking up the codebook, highlighting, underlining, inserting tabs, and making notes where appropriate. All of these fundamentals must be solidly in place before the Codeology method is applied in the Introduction and nine chapters of the *NEC*®.

REVIEW QUESTIONS

1. The *National Electrical Code*® user can customize his/her codebook by which of the following methods?
   a. Highlighting
   b. Making notes and underlining
   c. Use of code tabs
   d. All of the above

2. Using the Codeology method, the Code user must determine the proper clues or key words and go to the _____.
   a. most likely article
   b. table of contents
   c. index
   d. highlighted areas

3. After using key words or clues to get into the correct chapter of the *NEC*®, the Codeology user further qualifies his/her needs and gets into the correct _____.
   a. section
   b. part
   c. article
   d. subdivision

4. After using key words or clues to get into the correct chapter and then the correct article of the *NEC*®, the Codeology user further qualifies his/her needs and gets into the correct

   _____.
   a. section of the part
   b. part of the article
   c. table
   d. subdivision

5. Which Chapter of the NEC is considered the "loner"?
   a. 7
   b. 8
   c. 9
   d. 10

6. An article in the 600-series will deal with special _____.
   a. permission
   b. occupancies
   c. equipment
   d. none of the above

7. The Codeology Method will allow the code user to become _____ with the NEC.
   a. accurate
   b. fast
   c. confident
   d. all of the above

8. The Codeology Method requires that all Articles be highlighted in green and all Parts be highlighted in _____ for fast identification when using the NEC.
   a. yellow
   b. pink
   c. blue
   d. orange

Article 90, The Introduction to the NEC®

O U T L I N E

OBJECTIVES

After completing this unit, you should be able to:

1. Recognize that Article 90 contains directions essential for the Code user to understand for proper application of the *NEC®*
2. Understand that the ground rules contained in Article 90 govern the entire *NEC®*
3. Recognize that the purpose, scope, and arrangement of the *NEC®* are governed by Article 90

OVERVIEW

Article 90, the Introduction to the *NEC®*, stands alone before the following nine chapters as instructions to be read and understood before the use of this Code. Any equipment or appliance purchased by a consumer arrives with a detailed set of instructions. In order for the consumer to use the appliance or equipment and achieve the desired results, the instructions must first be read and understood.

NEC® HOW TO BEGIN, ARTICLE 90 INTRODUCTION

At the beginning of Article 90 it would be helpful for all users of the *NEC®* if the following warning were posted as shown in Figure 7–1.

FIGURE 7–1   Warning sign.

WARNING

**READ AND UNDERSTAND THESE INSTRUCTIONS
(Article 90) BEFORE ATTEMPTING TO USE THIS CODE.
FAILURE TO READ AND UNDERSTAND THESE INSTRUCTIONS
WILL RESULT IN THE FOLLOWING:**

- Lack of understanding of the purpose of the NEC

- Attempts to enforce the NEC where it does not apply

- Lack of enforcement where the NEC does apply

- Lack of understanding of how the NEC provisions relate to the factory installed wiring of electrical equipment

- Misapplication due to a lack of understanding of the arrangement of the NEC

- Misapplication due to a lack of understanding of the use of mandatory/permissive rules and explanatory material

- Misapplication of the NEC as a design manual

- Misapplication of the NEC as an instruction manual

- Misapplication due to a lack of understanding that the rules in the NEC are minimum standards

Before attempting to apply the provisions contained within the *NEC®* the Code user must read and understand the instructions, Article 90. Table 7–1 reviews the section numbers and titles for Article 90.

TABLE 7–1	Layout of *NEC*®, Article 90 Introduction

NEC® Title:	Introduction
Codeology Title:	The Ground Rules
Article Scope:	The Introduction and Ground Rules for the *NEC*®

Section	Section Title
90.1	Purpose
90.2	Scope
90.3	Code Arrangement
90.4	Enforcement
90.5	Mandatory Rules, Permissive Rules, and Explanatory Material
90.6	Formal Interpretations
90.7	Examination of Equipment for Safety
90.8	Wiring Planning
90.9	Units of Measurement

DidYouKnow?

The *NEC*® is not intended as a design guide or as an instruction manual for untrained persons.

ARTICLE 90 INTRODUCTION

The following is a review of the instructions to the *NEC*®, which are the nine sections contained in Article 90.

90.1 Introduction: What is the purpose of the *NEC*®?

Safety is the primary purpose of the *NEC*®. These *safety*-driven requirements are intended to protect *persons* and *property* through the standardization of safe installation practices.

90.1 Purpose (A) Practical Safeguarding

The purpose of this Code is the practical safeguarding of persons and property from hazards arising from the use of electricity.

90.1 Purpose (B) Adequacy

This text explains that the *NEC*® contains *minimum* requirements resulting in a *safe* installation. This section also explains that the use of the *NEC*® does not guarantee an efficient, convenient, expandable, or adequate installation. These are design issues and must be considered in addition to the minimum installation requirements of the *NEC*®.

90.1 Purpose (C) Intention

The *NEC*® is not intended as a design specification or an instruction manual for untrained persons. The *NEC*® is an installation tool designed and revised every three years for use by trained persons in the electrical industry, inspectors, engineers, and manufacturers of electrical equipment.

90.1 Purpose (D) Relation to Other International Standards

The *NEC*® is an international electrical installation standard. This text is intended to explain the relationship between the *NEC*® and another electrical code used internationally outside of the United States called Section 131 of International Electrotechnical Commission Standard 60364–1, Electrical Installations of Buildings. In essence the *NEC*® addresses all of the protection requirements contained in the IEC code.

90.2 Scope: Where does the *NEC*® apply?

Section 90.2 sets the stage for where the *NEC*® applies in three first-level subdivisions as follows:

90.2 Scope (A) Covered

The *NEC*® covers the installation of the following:
- Electrical conductors
- Electric equipment
- Raceways
- Signaling and communications conductors equipment and raceways
- Optical fiber cables and raceways

  The *NEC*® covers the following premises and structures on the load side of the service point, see Figures 7–2 a and b:
- Public and private premises, including buildings, structures, mobile homes, recreational vehicles, and floating buildings
- Yards, lots, parking lots, carnivals, and industrial substations
- Installations of conductors and equipment that connect to the supply of electricity
- Installations used by the electric utility, such as office buildings, warehouses, garages, machine shops, and recreational buildings, that are not an integral part of a generating plant, substation, or control center

FIGURE 7–2a This sketch depicts a dwelling supplied with an underground service. The *NEC*® does not apply to utility-owned service conductors. The *NEC*® applies to all of the electrical installation beyond the service point. The service point, noted in this sketch, is the point where the utility-owned conductors meet the premises wiring.

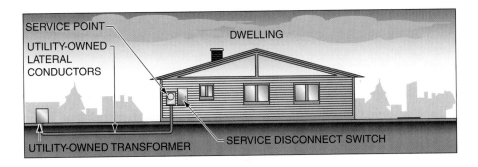

SERVICE POINT
UTILITY-OWNED LATERAL CONDUCTORS
DWELLING
UTILITY-OWNED TRANSFORMER
SERVICE DISCONNECT SWITCH

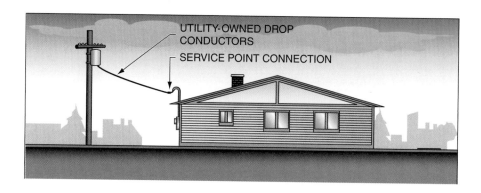

FIGURE 7–2b This sketch depicts a dwelling supplied with an overhead service. The NEC® does not apply to utility-owned service conductors. The NEC® applies to all of the electrical installation beyond the service point. The service point, noted in this sketch, is the point where the utility-owned conductors meet the premises wiring.

DidYouKnow?

The NEC® applies only to electrical installations as outlined in 90.2(A). The installations listed in 90.2 (B) are not covered by the NEC®.

90.2 Scope (B) Not Covered

The NEC® does not cover the following premises and structures:
*   Installations in ships, watercraft other than floating buildings, railway rolling stock, aircraft, or automotive vehicles other than mobile homes and recreational vehicles
*   Installations underground in mines and self-propelled mobile surface mining machinery and its attendant electrical trailing cable
*   Installations of railways for generation, transformation, transmission, or distribution of power used exclusively for operation of rolling stock or installations used exclusively for signaling and communications purposes
*   Installations of communications equipment under the exclusive control of communications utilities located outdoors or in building spaces used exclusively for such installations
*   Installations under the exclusive control of an electric utility where such installations:
    a. Consist of service drops or service laterals, and associated metering, or
    b. Are located in legally established easements or rights-of-way, either designated by or recognized by public service commissions, utility commissions, or other regulatory agencies having jurisdiction for such installations, or
    c. Are on property owned or leased by the electric utility for the purpose of communications, metering, generation, control, transformation, transmission, or distribution of electric energy. See Figures 7–3 a and b.

90.2 Scope (C) Special Permission

The AHJ (authority having jurisdiction) as defined in Article 100 may grant special permission when conductors used to connect to the utility are not under the control of the utility, provided the installation is located outdoors or terminates immediately inside a building.

FIGURE 7–3 Utility-owned substations and distribution are not covered by the *NEC®*.

(a)

(b)

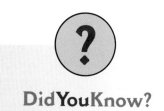

DidYouKnow?

The arrangement of the *NEC*® is outlined in Article 90.

90.3 Code Arrangement: How do these chapters apply?

The *NEC*® is arranged by chapter as follows:

General

Chapters 1 through 4 apply generally. This means they apply all of the time for every installation, in every occupancy unless supplemented or modified by Chapters 5, 6, or 7.

Special

Chapters 5, 6, and 7 address special requirements and supplement or modify Chapters 1 through 4.

Communications Systems

Chapter 8 stands alone. The remainder of the *NEC*® does not apply to articles within Chapter 8 unless specifically referenced in Chapter 8.

Tables

Tables located in Chapter 9 are applicable when they are referenced in other chapters.

Annexes

Annexes A through H are for informational purposes only.

90.4 Enforcement: How is this code to be enforced?

The *NEC*® is intended to be adopted by and enforced by governmental bodies exercising legal jurisdiction over electrical installations, such as a city, state, township, county, or municipality. The authority having jurisdiction (AHJ), as defined in Article 100, is responsible for making interpretations and may give special permission for specific installations when it is determined that an equivalent level of installation safety can be achieved (Figure 7–4).

90.5 Mandatory Rules, Permissive Rules, and Explanatory Material: When are the requirements of the *NEC*® mandatory?

The *NEC*® contains mandatory and permissive rules along with explanatory material, which is included to aid the user of this Code in understanding and applying the *NEC*®. Section 90.5 provides clear instructions to the Code user to determine which rules are mandatory or permissive as well as the identification of material designed to be informational only.

90.5(A) Mandatory Rules

A mandatory requirement exists when the terms *shall* or *shall not* are used within the *NEC*®.

FIGURE 7–4

When the *NEC*® is adopted and enforced, the authority having jurisdiction (AHJ) will require inspections of all electrical installations.

90.5(B) Permissive Rules

A permissive requirement exists when the terms *shall be permitted* or *shall not be required* are used within the *NEC®*.

90.5(C) Explanatory Material

Explanatory material is included in the *NEC®* in the form of *Fine Print Notes* that are always preceded by the acronym "FPN" and are written in small text below the section, subdivision, list item, or article to which they apply. FPNs are not an enforceable part of the *NEC®*.

90.6 Formal Interpretations: Is there a process for formal interpretations or is the local AHJ the ultimate authority?

Formal interpretation procedures are established in the NFPA Regulations Governing Committee Projects. This would require a detailed written request for a formal interpretation through NFPA. This process is in place to promote uniform interpretation and application of the *NEC®* in every city, state, township, county, or municipality in which it is adopted and enforced.

90.7 Examination of Equipment for Safety: Does the *NEC®* require that all listed factory-installed internal wiring of electrical equipment be inspected upon installation?

The *NEC®* does not intend that factory-installed internal wiring of listed equipment be inspected when installed. When qualified electrical testing laboratories have determined that equipment meets the appropriate product standard and they list and label the equipment accordingly, an inspection is not required (Figure 7–5).

90.8 Wiring Planning: Does the *NEC®* provide requirements for the future expansion of an electrical installation?

The *NEC®* does not require that an electrical installation be designed and constructed to allow for future expansion. The *NEC®* contains minimum safety-driven electrical installation requirements, not design elements for future expansion. Section 90.8 provides informational text that explains that electrical installations, designed with future expansion as a consideration, allow for efficient, convenient, and safe expansion (Figure 7–6).

90.8(A) Future Expansion and Convenience

Design considerations that are essential for convenient future expansion of electrical installations include plans and specifications that require:
* Ample space in raceways
* Spare raceways
* Additional space for electric power and communications circuits
* Readily accessible located distribution centers

DidYouKnow?

90.8 clearly illustrates that the NEC contains minimum installation requirements. Future expansion of the electrical system is a design issue and must be considered in addition to the minimum *NEC®* requirements.

FIGURE 7–5   The *NEC*® does not require that factory-installed internal wiring of equipment be inspected when installed.

90.8(B) Number of Circuits in Enclosures

This text explains that throughout the *NEC*® the number of circuits permitted in enclosures of all types is restricted to varying degrees dependent upon the installation and type of enclosure. This restriction of the number of circuits permitted exists to help minimize the damage that could occur to all circuits contained in an enclosure when one of those circuits experiences a ground fault or short circuit.

90.9 Units of Measurement: Why does the *NEC*® use both inch-pound and metric units of measurement?

When the *NEC*® specifies distance, sizes, or weight, both metric and inch-pound units are referenced because the *NEC*® is an international electrical installation standard. The United States still uses the inch-pound system while the rest of the world uses the metric system.

DidYouKnow?

The *NEC*® is an international document and contains both inch-pound and metric units to accommodate its use in the U.S. and internationally.

FIGURE 7-6 The *NEC®* does not require that an electrical installation be constructed with design considerations for future expansion.

90.9(A) Measurement System of Preference

Metric units used in the *NEC®* are in accordance with the modernized metric system known as the International System of Units (SI).

90.9(B) Dual System of Units

Both inch-pound and metric units are referenced together when the *NEC®* requires a specific distance, size, or weight. During the code-making cycle that produced the 2002 *NEC®*, metric units were recognized as the international standard and all metric units were placed first with the inch-pound units following in parentheses. For example, a cable assembly or other wiring method may require support every 1.8 m (6 ft.) and within 300 mm (12 in.) of all terminations.

90.9(C) Permitted Uses of Soft Conversions

This section gives instructions on how conversions from inch-pound to metric and vice versa are permitted to be made. An exact conversion from inch-pound to metric is known as a *soft* conversion. When a requirement within the *NEC*® directly impacts safety, such as working space in 110.26(A), a *soft* or exact conversion is required. A conversion that is permitted to be approximate or rounded off to the nearest standard unit is called a *hard* conversion. For standardization of requirements for support of wiring methods, *hard* or rounded off conversions are used.

90.9(D) Compliance

Approximate or *hard* conversions are permitted when they do not have a negative impact on safety. Compliance with either the metric or the inch-pound system meets the requirements of the *NEC*®. The choice of which units to use, metric or inch-pound, is determined by the user of this Code.

Notations Used in the *NEC*®

Notations that are used to help the code user identify changes from the previous edition of the *NEC*® are not explained in Article 90. This editorial information is handled directly by NFPA staff and is provided on the inside cover of the *NEC*®.

These notations are as follows:

1. A vertical line is placed next to a technical change that has occurred where:
   a. the change occurs in large blocks.
   b. large areas of new text are added.
   c. the change occurs in tables and figures.
   d. new tables or figures are added.
2. Changes other than editorial are highlighted with gray shading within sections.
3. Where one or more complete paragraphs are deleted a bullet • is placed in the margin between the remaining paragraphs or sections, etc.

DidYouKnow?

90.9(D) allows the Code user to apply either the metric or inch-pound systems, to comply with requirements in the *NEC*®.

SUMMARY

The nine sections contained in Article 90 provide the ground rules for the use, application, and scope of the entire *National Electric Code®*. The title of Article 90 is "Introduction." This article introduces the ground rules or directions for the use of the *NEC®*. The nine sections contained in this article are as follows:

90.1	Purpose
90.2	Scope
90.3	Code Arrangement
90.4	Enforcement

90.5	Mandatory Rules, Permissive Rules, and Explanatory Material
90.6	Formal Interpretations
90.7	Examination of Equipment for Safety
90.8	Wiring Planning
90.9	Units of Measurement

This Introduction must be understood prior to the use of the rules or information contained in the *NEC®*. These are the directions given to the users of this Code for proper understanding and application of the *NEC®*.

REVIEW QUESTIONS

1. The primary purpose of the *National Electrical Code®* is _____.
   a. cost efficient installations
   b. to increase installation cost
   c. safety
   d. training

2. The *NEC®* contains _____ for electrical installations.
   a. provisions for convenient expansion
   b. training of persons
   c. a design manual
   d. minimum requirements

3. The *NEC®* is not intended for use as _____.
   a. an instruction manual
   b. a tool for untrained persons
   c. a design manual
   d. all of the above

4. The use of mandatory rules, permissive rules, and explanatory material throughout the *NEC®* is governed by Article 90 in section _____.
   a. 90.1
   b. 90.3
   c. 90.2
   d. 90.5

5. The use of the term "shall" represents _____ text in the *NEC®*.
   a. mandatory
   b. permissive
   c. informational
   d. none of the above

6. The use of the phrase "shall be permitted" represents _____ text in the *NEC®*.
   a. mandatory
   b. optional
   c. informational
   d. none of the above

7. The use of a Fine Print Note represents _____ text in the *NEC®*.
   a. mandatory
   b. optional
   c. informational
   d. none of the above

8. The Code user is permitted by _____ to use metric or inch-pound units to be in compliance with the *NEC®*.
   a. 90.5(B)
   b. 90.2(C)
   c. 90.9(B)
   d. 90.9(D)

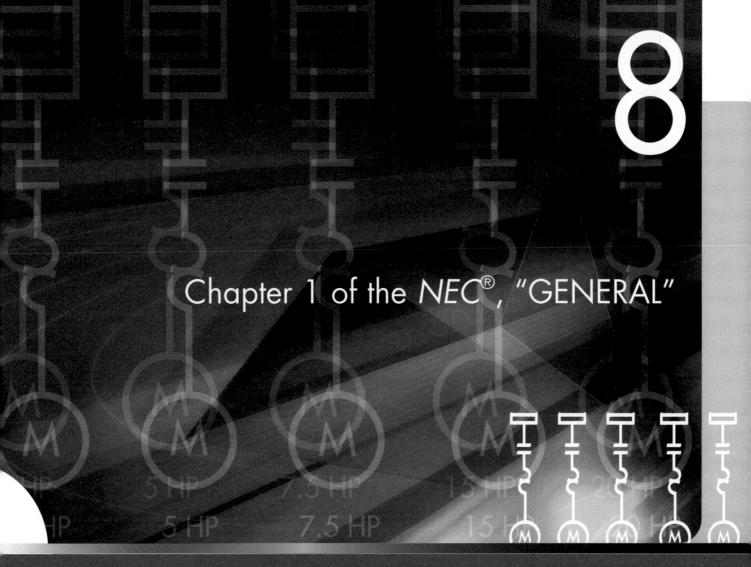

Chapter 1 of the NEC®, "GENERAL"

8

OBJECTIVES

After completing this unit, you should be able to:

1. Associate the Codeology title for *NEC*® Chapter 1 as "General"
2. Describe the general type of information and requirements contained in Chapter 1
3. Recognize key words and clues for locating answers in *NEC*® Chapter 1, the "General" chapter
4. Identify exact sections, subdivisions, and list items, etc. to justify answers for all questions referring to Chapter 1 articles
5. Identify that Chapter 1 numbering is the 100 series
6. Recognize, recall, and become familiar with articles contained in Chapter 1

OVERVIEW

The Codeology title for Chapter 1 is "General." All electrical installations will be subject to general rules that will apply in all electrical installations. For example, definitions existing in Article 100, will apply to the terms defined where ever they are used in this Code. General requirements also exist in the "General" chapter in Article 110 covering basic needs that are common to all electrical installations. In accordance with the Code arrangement requirements of 90.3, Chapter 1 applies to all electrical installations unless supplemented or modified in Chapters 5, 6, or 7.

NEC®: CHAPTER 1—GENERAL

There are two articles in Chapter 1, the 100-series. These two articles provide general requirements for all electrical installations and definitions essential for the proper application of the *NEC®*.

Article 100 Definitions

Article 100 contains only those definitions essential to the proper application of this Code. It is not intended to include commonly defined general terms or commonly defined technical terms from related codes and standards. In general, only those terms that are used in two or more articles are defined in Article 100. Other definitions are included in the article in which they are used but may be referenced in Article 100. Table 8–1 reviews the articles for Chapter 1.

TABLE 8–1   Layout of *NEC®*, Chapter 1

NEC® Title: General	
Codeology Title: General	
Chapter Scope: General Information and Rules for Electrical Installations	
Article	Title
100	Definitions
110	Requirements for Electrical Installations

The *NEC®* does not attempt to define commonly used general terms or commonly used and defined technical terms. Article 100 contains definitions for terms that are not common or are essential to the proper application of the *NEC®*. A term that needs to be defined in the *NEC®* is placed in Article 100 only if the term being defined is used in two or more articles. Terms that are defined in the *NEC®* and are used in only one article are defined in the second section of the article in which they exist. For example, in "Article 517 Health Care Facilities" there are 39 definitions in 517.2, the second section of the article. When terms are defined within an individual article, they apply only within that article.

Part I of Article 100 contains definitions intended to apply wherever the terms are used throughout this Code. In addition, Part I contains 158 definitions that apply to the defined terms wherever they are used in the *NEC®*. Part II contains definitions applicable only to the parts of articles specifically covering installations and equipment operating at over 600 volts, nominal. Part II contains four definitions that apply only to parts of articles specifically covering installations and equipment operating at over 600 volts, nominal. For example, terms defined in Part II of Article 100 would apply to Part IX of Article 240.

Misapplication of the requirements of the *NEC®* occurs regularly due to a lack of understanding of terms defined in the *NEC®*. For example, in order to understand and apply the requirements of "Article 230 Services," the user of this Code must understand the terms defined in Article 100, including but not limited to the following (see also Figures 8–1a and b):

- **Service.** The conductors and equipment for delivering electric energy from the serving utility to the wiring system of the premises served.
- **Service Cable.** Service conductors made up in the form of a cable.
- **Service Conductors.** The conductors from the service point to the service disconnecting means.
- **Service Drop.** The overhead service conductors from the last pole or other aerial support to and including the splices, if any, connecting to the service-entrance conductors at the building or other structure.
- **Service-Entrance Conductors, Overhead System.** The service conductors between the terminals of the service equipment and a point usually outside the building, clear of building walls, where joined by tap or splice to the service drop.
- **Service-Entrance Conductors, Underground System.** The service conductors between the terminals of the service equipment and the point of connection to the service lateral.
  - FPN: Where service equipment is located outside the building walls, there may be no service-entrance conductors, or they may be entirely outside the building.
- **Service Equipment.** The necessary equipment, usually consisting of a circuit breaker(s) or switch(es) and fuse(s) and their accessories, connected to the load end of service conductors to a building or other structure, or an otherwise designated area, and intended to constitute the main control and cutoff of the supply.
- **Service Lateral.** The underground service conductors between the street main, including any risers at a pole or other structure or from transformers, and the first point of connection to the service-entrance conductors in a terminal box or meter or other enclosure, inside or outside the building wall. Where there is no terminal box, meter, or other enclosure, the point of connection is considered to be the point of entrance of the service conductors into the building.
- **Service Point.** The point of connection between the facilities of the serving utility and the premises wiring.

There are many other definitions contained in Article 100 that are essential to the proper application of Article 230. However, there are nine different definitions needed to properly identify conductors and equipment to ensure proper application of Article 230. Before interpreting and applying the rules of Article 230 or any other Article the Code user must understand the definitions which apply to ensure proper understanding and application.

ARTICLE 110 REQUIREMENTS FOR ALL ELECTRICAL INSTALLATIONS

Article 110 contains general requirements that are common to all electrical installations. Article 110 is separated into five parts to address general requirements for all electrical installations. The following key provisions of

DidYouKnow?

For proper application of the *NEC®*, it is essential that the Code user be familiar with and understand definitions in Article 100 and those contained in individual articles.

FIGURE 8-1a-b The definitions located in Article 100 apply to all electrical installations, including services, and to all articles in Chapters 1 through 7.

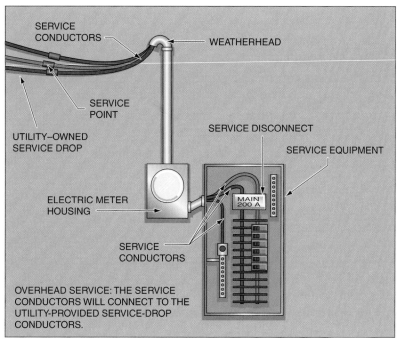

(a)

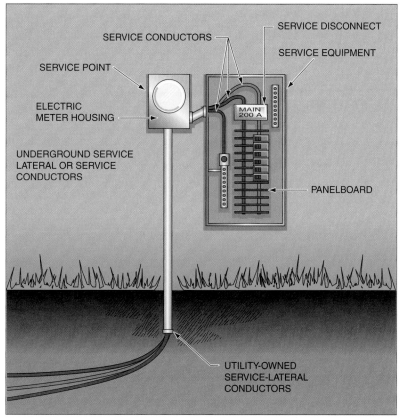

(b)

this article represent the general nature of these requirements as they apply to all electrical installations.

PART I GENERAL

110.1 Scope

These general requirements for all electrical installations include but are not limited to the following:
- Examination and approval of conductors and equipment
- Installation and use of conductors and equipment
- Access to electrical equipment
- Spaces about electrical equipment
- Enclosures intended for personnel entry
- Tunnel installations

110.2 Approval

This section requires that all conductors and equipment be *approved* as defined in Article 100. Approval occurs when the conductors and equipment are acceptable to the AHJ, the authority having jurisdiction.

110.3 Examination, Identification, Installation, and Use of Equipment

110.3(A) Examination

This text requires that equipment be examined to evaluate suitability, strength, durability, bending space, connection space, heating, arcing, and classification by type, size, voltage, current capacity and use.

110.3(B) Installation and Use

Listed or labeled equipment is required to be installed in accordance with the listing and labeling of the equipment (see Figures in 8–2a–e). For example, a conductor termination point labeled "cu" is permitted only for copper conductors while the labeling of "al/cu" on a termination point permits copper or aluminum conductors.

110.5 Conductors

Conductors referenced in the *NEC*® are copper unless otherwise noted.

110.6 Conductor Sizes

Conductor sizes used in the *NEC*® are expressed in the American Wire Gauge (AWG) or in circular mils. The AWG uses numbers, for example, 12 AWG and 1/0 AWG. When circular mils are used, the *NEC*® commonly uses kcmil or 1,000 mils, for example, 500 kcmil.

DidYouKnow?

110.3(B) requires that listed or labeled equipment be installed in accordance with instructions included in the listing or labeling.

DidYouKnow?

110.9 requires that equipment intended to interrupt current at fault levels have an interrupting rating sufficient for the current to be interrupted.

Part I General 111

FIGURE 8-2a-e The *NEC®* requires that listed and labeled equipment be installed in accordance with both the equipment listing requirements and any labeling on the equipment.

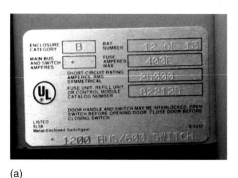

(a)

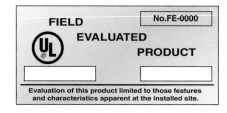

(b)

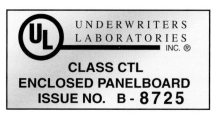

(c)

(d)

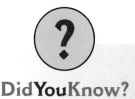

(e)

110.9 Interrupting Rating

All equipment intended to interrupt current at fault levels must have a sufficient interrupting rating.

110.10 Circuit Impedance and Other Characteristics

Damage to electrical systems from short circuits and ground faults must be minimized by the proper selection of protective equipment and the consideration of all characteristics of the equipment and conductors protected.

110.12 Mechanical Execution of Work

This section requires that all electrical equipment be installed in a neat and workmanlike manner. Electrical installations must be installed neatly and with care to be considered to be installed in a skillfully or in a workmanlike manner.

DidYouKnow?

110.12 requires that electrical equipment be installed in a "neat and workmanlike manner." This means a skillful and neat-looking installation.

110.12(A) Unused Openings

All unused cable or raceway openings in all electrical equipment must be effectively closed.

110.12(B) Integrity of Electrical Equipment and Connections

All electrical equipment must be kept clean and protected from damage, corrosion, overheating, or deterioration.

110.13 Mounting and Cooling of Equipment

110.13(A) Mounting

All electrical equipment must be firmly secured to the surface on which it is mounted. Wooden plugs are prohibited to be used to mount or secure electrical equipment.

110.13(B) Cooling

Electrical equipment that depends on the natural circulation of air for cooling purposes must be installed so that other equipment, walls, or other structural members do not block or limit the natural circulation of air. Electrical equipment, such as a transformer, will generate heat during normal operation. This section requires that all electrical equipment that may generate heat during normal operation be installed to allow for cooling through natural circulation of air.

110.14 Electrical Connections

All electrical installations will contain electrical connections. This section contains general requirements for all electrical connections in all electrical installations. These requirements are as follows:
- Due to the different characteristics of conductor metals, such as aluminum and copper, all terminals, connectors, and lugs shall be identified (labeled/marked) for the conductor material permitted.
- Dissimilar metals, such as aluminum and copper, shall not be intermixed in a connector unless the connector is identified or listed for the purpose.
- Solder, flux, inhibitors, and conductor compounds must not have an adverse effect on the conductors or equipment.

110.14(A) Terminals

Conductors must be terminated in a manner that ensures a solid connection without damaging the conductors. Terminals designed for the termination of more than one conductor must be identified for such use. Terminals designed for use with aluminum conductors must also be identified for the purpose.

110.14(B) Splices

Conductors shall be spliced by one of the following methods:
- Devices identified for the purpose (i.e., wire nuts)
- Brazing/welding/soldering with a fusible metal or alloy
  - Soldered splices must first be mechanically joined

- All splices and free ends of conductors must be insulated
- Wire nuts or splicing devices for direct burial shall be listed for the purpose

110.14(C) Temperature Limitations

All electrical terminations and conductors have temperature limitations. Typical temperature limitations are 60°C, 75°C, and 90°C. The intent of this requirement is to ensure that the lowest temperature rating in an electrical system or circuit is not exceeded. For example, a circuit may be fed from an over current device rated at 75°C using conductors rated at 90°C and terminate in equipment rated at 60°C. The lowest temperature rating of 60°C must then be applied. Derating of the conductor, if required, is permitted to start at the higher 90°C rating.

110.15 High-Leg Marking

The use of a 3-phase, 4-wire delta connected system results in one phase having a higher voltage to ground. For example, a 120/240-volt, 3-phase, 4-wire system is required, in Article 250 of the *NEC®*, to be grounded. The voltage to ground from "A" phase/leg is 120-volts. The voltage to ground from "C" phase/leg is 120-volts. The "B" phase/leg voltage to ground is significantly higher at 208-volts to ground. This section requires that the *high leg* be identified with the color orange wherever connections are made and both the grounded conductor and the high leg are available. This marking is required to prevent misapplication of the high leg.

110.16 Flash Protection

Electrical hazards include shock hazards, arc flash hazards, and arc blast hazards. An arc flash can produce temperatures of 35,000°F at the point of contact and temperatures in the ambient space of an electrical worker can easily reach upwards of 15,000°F. The result of an accident involving an arc flash can result in serious injuries from incurable third-degree burns and death.

This marking requirement is designed to warn qualified persons of potential arc flash hazards (see sample marking in Figure 8–3a-b). Arc flash warnings are required to be marked on all switchboards, panelboards, industrial control panels, meter socket enclosures, and motor control centers in other than dwelling units. An informational FPN informs the user of this Code that NFPA-70E, the "Standard for Electrical Safety in the Workplace," provides assistance in determining the severity of an exposure, safe work practices, and required personal protective equipment.

110.21 Marking

All electrical equipment is required to be durably marked with the manufacturer's name, trademark, or other distinctive marking for identification purposes.

110.22 Identification of Disconnecting Means

All disconnecting means are required to be legibly marked to identify their purpose, unless the installation is arranged so that the purpose of the disconnecting means is clearly evident.

DidYouKnow?

110.14 provides general requirements for electrical connections in all electrical installations.

DidYouKnow?

110.16 provides a general requirement for marking all switchboards, panelboards, industrial control panels, meter-socket enclosures, and motor control centers in other than dwelling units with an "Arc Flash Warning."

FIGURE 8-3 General marking requirements such as "Arc Flash Warnings" are located in Chapter 1 of the *NEC*®.

(a)

(b)

? DidYouKnow?

Requirements in Part II of Article 110 apply generally in all installations to equipment rated at 600 volts or less.

PART II 600 VOLTS, NOMINAL OR LESS

110.26 Spaces About Electrical Equipment

This section requires sufficient access to equipment and working space be maintained (see Figure 8–4).

All electrical equipment likely to require examination, adjustment, servicing, or maintenance while energized must provide adequate working space as required in 110.26(A)(1), (2) & (3).

FIGURE 8-4   General requirements such as "Working Space" around electrical equipment are located in Chapter 1 of the *NEC*®.

110.26(A) Working Space

(1) Depth of working space Adequate working space is required for the safety of electrical workers who install and maintain electrical systems. Table 110.26(A)(1) Working Spaces specifies the minimum working space in front of electrical equipment. This table separates systems 600-volts or less into two categories using the system voltage to ground. The first is systems from 0 to 150 volts to ground and the second is systems from 151 to 600 volts to ground. Minimum clear distance for working space is then determined from one of three conditions:

* Condition 1: when the equipment is opposite no other electrical equipment or any grounded objects or parts.
* Condition 2: when the equipment is opposite grounded objects or parts. Note that concrete, brick, or tile walls are considered grounded.
* Condition 3: when the equipment is opposite other electrical equipment.

(2) Width of working space The width of working space in front of all electrical equipment shall be the width of the equipment or 30 inches,

DidYouKnow?

110.26 provides general working space requirements for all electrical installations.

whichever is greater. In all cases, hinged panels and doors must be capable of opening a full 90 degrees.

(3) Height of working space The working space shall be clear and extend from the floor or working platform to the height of the equipment or $6\frac{1}{2}$ feet, whichever is greater.

110.26(B) Clear Spaces

The working space in front of electrical equipment shall not be used for storage.

110.26(C) Entrance to Working Space

(1) Minimum required In general, a minimum of one entrance is required for access to working space about electrical equipment.

(2) Large equipment In general, when equipment rated at 1200 amps or more and over 6 feet wide containing overcurrent, switching, or control devices exists, there shall be one entrance at each end of the working space not less than $6\frac{1}{2}$ feet high and 24 inches wide. However, provisions exist to permit a single extrance to these spaces where the exit is unobstructed or additional working space is provided.

(3) Personnel Doors Where equipment rated 1200 amps or more containing overcurrent, switching, or control devices is located, all personnel doors less than 25 feet from the nearest edge of the working space must open in the direction of egress and be equipped with panic-type hardware to permit a quick exit in the event of an electrical fault.

110.26(D) Illumination

Working spaces around service equipment, switchboards, panelboards, or motor control centers installed indoors require illumination to provide electrical workers maintaining the system with adequate lighting to perform routine tasks.

110.26(E) Headroom

The minimum amount of headroom is determined by the height of the equipment or $6\frac{1}{2}$ feet, whichever is greater. In existing dwelling units for service equipment rated 200 amps or less, the headroom is permitted to be less than $6\frac{1}{2}$ feet.

110.26(F) Dedicated Equipment Space

DidYouKnow?

110.26(F) provides general requirements for dedicated equipment space for all electrical installations.

This section applies to all switchboards, panelboards, distribution boards, and motor control centers and requires that where they are installed, the locations are dedicated to this equipment. This requirement is not working space for the electrical worker, it is dedicated equipment space to allow for raceways and cable assemblies to enter equipment. This requirement is also intended to prevent foreign systems from being installed above electrical equipment. The dedicated space addressed in this section is the area above the footprint formed by the top of the equipment (see Figure 8–5). In gen-

FIGURE 8-5 General requirements such as "Dedicated Equipment Space" above electrical equipment are located in Chapter 1 of the *NEC*®.

eral, the area 6 feet above the footprint formed by the top of the equipment or the structural ceiling, whichever is lower, is dedicated equipment space.

PART III OVER 600 VOLTS, NOMINAL

Part III of Article 110 is dedicated to general requirements for all electrical installations for systems rated at over 600 volts. Section 110.30 clearly explains that the requirements of Part III are in addition to the general provisions of Part I of Article 110. This section also states for clarity that Part III does not apply on the supply side or upstream of the service point. The general requirements in Part III address areas very similar to Part II of Article 110. The difference is that the requirements of Part III are modified to address the safety of high-voltage system installations.

PART IV TUNNEL INSTALLATIONS OVER 600 VOLTS, NOMINAL

Part IV of Article 110 is dedicated to tunnel installations with systems and circuits rated over 600 volts. Section 110.51 explains in detail the areas covered by Part IV. This part covers the installation and use of high-voltage power distribution and utilization equipment that is portable, mobile, or both. Examples of these include but are not limited to substations, trailers, cars, mobile shovels, draglines, hoists, drills, dredges, compressors, pumps, conveyors, underground excavators, and the like.

DidYouKnow?

Article 110 is separated into five Parts. This logical separation of information and requirements allows the Code user following the Codeology Method to quickly and accurately locate needed information.

PART V MANHOLES AND OTHER ELECTRIC ENCLOSURES INTENDED FOR PERSONNEL ENTRY, ALL VOLTAGES

Part V of Article 110 is dedicated to manholes and other electric enclosures intended for personnel entry at all voltages. This part addresses working space inside these enclosures to provide adequate access and space, where equipment or parts contained are likely to require examination, adjustment, servicing, or maintenance while energized. Enclosures must also be of sufficient size to allow for the installation and removal of conductors without damaging the conductor insulation.

KEY WORDS/CLUES FOR CHAPTER 1—GENERAL

- General definitions for terms used in more than one article
- Determination of "approval"
- Examination of equipment
- Identification of equipment
- Installation and use of equipment
- General questions about electrical installations
- Mechanical execution of work
- Mounting and cooling of equipment
- Electrical connections
- High-leg marking requirements
- Flash protection
- Markings
- General identification of disconnects
- General electrical questions for over 600 volts
- General questions for manholes and tunnels
- Working space

SUMMARY

Chapter 1, in accordance with Section 90.3, applies generally to all electrical installations. Using the Codeology method, this chapter is given the nickname of the "General" chapter, due to the scope covered. The scope of this chapter can be described as "General Information and Rules for all Electrical Installations." Chapter 1 covers the entire electrical system from the service point (connection to the utility) to the last receptacle or other outlet in the electrical system.

See Figure 8–6 on page 120 for an example of how Chapter 1, along with Chapters 2, 3, and 4, build the foundation or backbone of all electrical installations.

REVIEW QUESTIONS

1. Does Chapter 1 of the *NEC®* apply to all electrical installations covered by the *NEC®*?

2. Chapter 1 is subdivided into how many articles?

3. Does Article 100 contain all terms defined in the *NEC®*?

4. When is a term defined in the *NEC®* placed in Article 100?

5. When a term is defined in an article other than Article 100, it is always placed in which section of the article?

6. Do the provisions of Part II of Article 100 apply to the installation of a 120/240-volt panelboard in a dwelling unit?

7. Workspace clearances are general requirements for all electrical installations. Where is this general requirement for equipment rated at 240 volts located in Chapter 1 of the *NEC®*?

8. How many parts is Article 110 subdivided into?

9. Is Part IV of Article 110 limited to installations above 600-volts?

10. Which part of Article 110 would contain the requirement that all listed and labeled equipment be installed in accordance with their listing and labeling?

FIGURE 8-6 Chapter 1 will apply generally in all electrical installations.

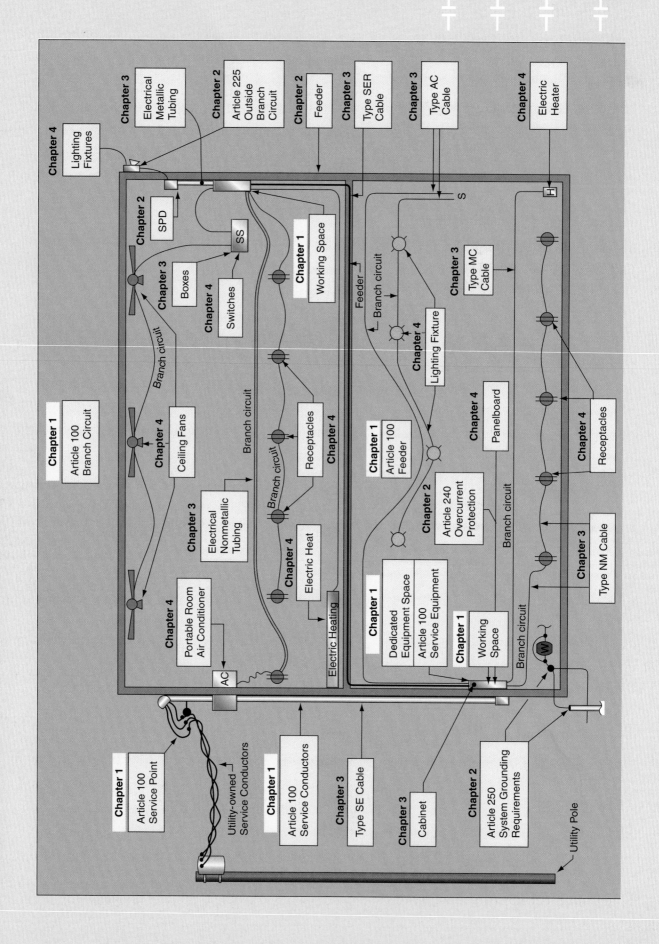

PRACTICE PROBLEMS

Using the Codeology Method

The following questions and steps to find the answer are designed to illustrate the Codeology method. While the steps to find your answer may seem lengthy, they are designed to illustrate the thought process to find the answer. The easiest way to make the Codeology method work is to silently "talk to yourself." Walk through these steps by silently talking to yourself and Codeology will be a natural response for quickly and accurately finding needed information in the *NEC*®.

Read the following questions and follow step by step using your codebook and the Codeology method.

1. A 120/208-volt panelboard being installed in a commercial occupancy is being installed in a small closet dedicated to electrical equipment. The minimum width of working space for the panelboard is _____ inches.

Step #1

Qualify the question or need, look for key words

This question is about working space for electrical equipment.

Key words: *"working space"*

Think general requirement, think General

Step #2

Go to the table of contents, Chapter 1, the General chapter

Find the correct article   **110 Requirements for Electrical Installations**

Step #3

Further qualify your question or need and get in the right part of the article

Keywords: *"working space for the panelboard"*

This question is about equipment installation under 600 volts.

Look in   **Part II 600 Volts, Nominal, or Less.**

Step #4

Read each section title until the correct section is found

Identify correct section or subdivision and find answer

Correct section/subdivision

Section   **110.26 Spaces About Electrical Equipment**

First-level subdivision   **(A) Working Space**

Second-level subdivision   **-(2) Width of Working Space**

2. Electrical equipment provided with ventilating openings shall be installed so that walls or other obstructions do not prevent the _____ _____ of air through the equipment to cool exposed surfaces.

Step #1

Qualify the question or need, look for key words

This question is general, about cooling of electrical equipment.

Keywords: *"air through the equipment"*

Think general requirement, think General

Step #2

Go to the table of contents, Chapter 1, the General chapter

Find the correct article   **110 Requirements for Electrical Installations**

Step #3

Further qualify your question or need and get in the right part of the article

Keywords: *"air through the equipment to cool exposed surfaces"*

This question is general for all electrical equipment.

Look in   **Part I General**

Step #4

Read each section title until the correct section is found

Identify correct section or subdivision and find answer

Correct section/subdivision

Section   **110.13 Mounting and Cooling of Equipment**

First-level subdivision   **(B) Cooling**

3. Round access openings in a manhole for personnel shall not be less than _____ inches in diameter.

Step #1

Qualify the question or need, look for key words

This question is general, about a manhole.

Keywords: *"access openings in a manhole"*

Think general requirement, think General

Step #2

Go to the table of contents, Chapter 1, the General chapter

Find the correct article   **110 Requirements for Electrical Installations**

Step #3

Further qualify your question or need and get in the right part of the article

Keywords: *"Round access openings in a manhole for personnel"*

This question is general for all electrical equipment.

Look in   **Part V Manholes and Other Electric Enclosures Intended for Personnel Entry, All Voltages**

Step #4

Read each section title until the correct section is found

Identify correct section or subdivision and find answer

Correct section/subdivision

Section   **110.75 Access to Manholes**

First-level subdivision   **(A) Dimensions**

4. The space above the top of a 277/480-volt switchboard installed indoors is dedicated equipment space for raceways and cable assemblies, from the top of the equipment for a distance of _____ feet or the structural ceiling whichever is lower.

Step #1

Qualify the question or need, look for key words

This question is general, about dedicated equipment space.

Keywords: *"dedicated equipment space"*

Think general requirement, think General

Step #2

Go to the table of contents, Chapter 1, the General chapter

Find the correct article   **110 Requirements for Electrical Installations**

Step #3

Further qualify your question or need and get in the right part of the article

Keywords: *"dedicated equipment . . . from the top of the equipment"*

This question is about equipment installation under 600-volts.

Look in   **Part II 600 Volts, Nominal, or Less**

Step #4

Read each section title until the correct section is found

Identify correct section or subdivision and find answer

Correct section/subdivision

Section   **110.26 Spaces About Electrical Equipment**

First-level subdivision   **(F) Dedicated Equipment Space**

Second-level subdivision   **(1) Indoor**

Third-level subdivision   **(a) Dedicated Electrical Space**

5. The *NEC®* requires that a "qualified person" have skills and knowledge related to the construction and operation of the electrical equipment and installations and has received _____ training to recognize and avoid the hazards involved.

Step #1

Qualify the question or need, look for key words

This question is general, about a "qualified person."

Keywords: *"qualified person"*

Think general requirement, think General

Step #2

Go to the table of contents, Chapter 1, the General chapter

Find the correct article   **100 Definitions**

Further qualify your question or need and get in the right part of the article

Keywords: *"qualified person"*

This question is about the definition of "qualified person."

Look in   **Part I General**

Read each section title until the correct section is found

Identify correct section or subdivision and find answer

Correct section/subdivision

Article   **100 Definitions**

Definition   **Qualified Person**

Chapter 2 of the NEC®, "PLAN"

O U T L I N E

WIRING

Article 200 Use and Identification of
Grounded Conductors
Article 210 Branch Circuits
Article 215 Feeders
Article 220 Branch Circuit, Feeder and
Service Calculations
Article 225 Outside Branch Circuits and
Feeders
Article 230 Services

PROTECTION

Article 240 Overcurrent Protection
Article 250 Grounding and Bonding
Article 280 Surge Arrestors, Over 1 kV
Article 285 Surge Protective Devices (SPDs),
1 kV or Less

KEY WORDS/CLUES FOR CHAPTER 2,
"PLAN"

THINK "PLAN" AND GO TO CHAPTER 2

OBJECTIVES

After completing this unit, you should be able to:
1. Associate the Codeology title for *NEC®* Chapter 2 as "Plan"
2. Identify the planning type of information and requirements for wiring and protection contained in Chapter 2
3. Recognize key words and clues for locating answers in *NEC®* Chapter 2, the Plan chapter
4. Identify exact sections, subdivisions, list items, etc. to justify answers for all questions referring to Chapter 2 articles
5. Recognize that Chapter 2 numbering is the 200-series
6. Recognize, recall, and become familiar with articles contained in Chapter 2

OVERVIEW

The Codeology title for Chapter 2 is "Plan." All electrical installations must be planned to suit many needs (Figure 9–1). These needs include but are not limited to the following:
- Design: the electrical installation must suit the specific needs of the occupancy, building, structure, or location in which it is installed
- Compliance with the *NEC®*
- Compliance with local building codes
- Plan submission to Authority Having Jurisdiction, the AHJ

The planning stage of all electrical installations includes two very important areas: wiring and protection.

FIGURE 9–1   All electrical installations must be properly planned before the building of the installation.

WIRING

In the planning stages of any electrical installation, the designer's intent is seen on the electrical drawings. These drawings will provide the installer with the necessary information to BUILD the installation. All electrical installations must have a source of electrical energy that is, in most cases, a service (Article 230) from a local utility company. From the service equipment, feeders (Article 215) will supply panelboards to distribute electrical energy throughout the building. From the panelboards, branch circuits (Article 210) will supply receptacle outlets, lighting fixtures, and other utilization equipment to facilitate the USE of electrical energy. Branch circuits and feeders installed outside of a building or structure must also comply with outside branch circuits and feeders (Article 225). All of the different types of conductors mentioned above must be properly sized to handle the intended load. This requires that calculations (Article 220) be applied to properly size each conductor. When the electrical system/s employ a grounded

Did You Know?

The three primary types of current-carrying conductors addressed in the *NEC*® are:
- Service conductors
- Feeder conductors
- Branch circuit conductors

conductor (Article 200), the installer must plan for the proper use and identification of these conductors.

Code compliance for current-carrying conductors must be planned into the electrical installation. This planning, covered in the wiring articles of Chapter 2, focuses only on the *NEC*® terms for the current-carrying conductors, not a type of raceway or cable assembly (Figure 9–2a and 9–2b). These *NEC*® terms for conductors are major clues for you to start in Chapter 2. The *NEC*® terms that deal with planning the wiring of an electrical installation include the following:

- Branch circuits
- Feeders
- Services
- Calculations, computed load/s
- Grounded conductors

FIGURE 9–2a The planning requirements in the wiring articles of Chapter 2 focus on the *NEC*® terms for current-carrying conductors.

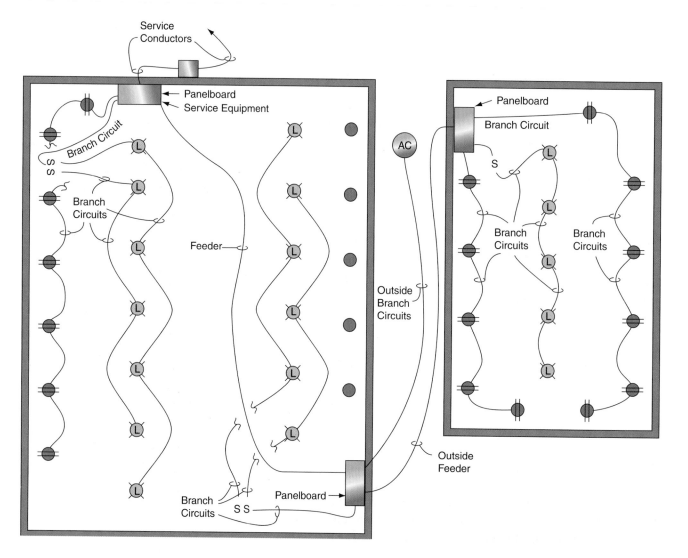

FIGURE 9–2b Chapter 2 of the NEC, the PLAN chapter, does not address specific wiring methods. This chapter uses the NEC terms service, feeder, and branch circuit.

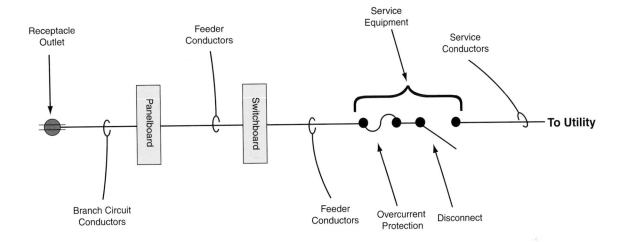

Chapter 2 of the NEC, the PLAN chapter, does not address specific wiring methods. This chapter uses the NEC terms service, feeder, and branch circuit.

Articles dedicated to wiring are shown in Table 9–1:

TABLE 9–1 Layout of *NEC®*, Chapter 2: Wiring and Protection

NEC® Title:	Wiring and Protection
Codeology Title:	Plan
Chapter Scope:	Information and Rules on Wiring and Protection of Electrical Installations

Wiring	
Article	Article Title
200	Use and Identification of Grounded Conductors
210	Branch Circuits
215	Feeders
220	Branch Circuit, Feeder and Service Calculations
225	Outside Branch Circuits and Feeders
230	Services

Article 200 Use and Identification of Grounded Conductors

The definition of the term *grounded conductor* exists in Article 100 and is defined as *a system or circuit conductor that is intentionally grounded* (Figure 9–3). Grounded conductors are often called the *neutral* conductor. It is extremely important to note that not all grounded conductors are neutrals. When a current-carrying conductor is common to all of the ungrounded (HOT) con-

DidYouKnow?

Not all "grounded conductors" are "neutral conductors."

A "grounded conductor" in a single-phase, three-wire system is also a "neutral conductor."

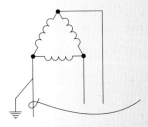

A "grounded conductor" in a three-phase three-wire system is a "phase conductor" and is not a "neutral conductor."

FIGURE 9–3   The *NEC*® defines all current-carrying and grounding conductors. The "ungrounded" conductor is commonly called the hot conductor. The "grounded" conductor is commonly called the neutral.

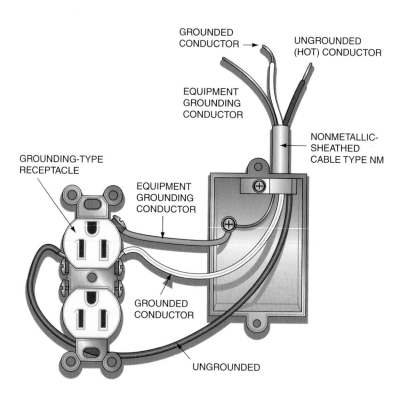

ductors of the system, it is considered neutral. When that *common* or neutral conductor is intentionally grounded, it becomes the grounded conductor.

Article 200 provides requirements for the following:

(1) Identification of terminals for grounded conductors

(2) Use of grounded conductors in premises wiring systems

(3) Identification of grounded conductors

Article 210 Branch Circuits

The definition of the term *branch circuit* exists in Article 100 and is defined as *the circuit conductors between the final overcurrent device protecting the circuit and the outlet(s)* (Figure 9–4).

This basic definition defines the current-carrying conductors that supply current to the load. There are many different types of branch circuits defined in Article 100. These definitions are designed to differentiate between the different uses of branch circuits and the specific rules within the *NEC*® for each type of branch circuit. Article 210 provides the general requirements for all branch circuits. Note that in the scope of this article that branch circuits for motor loads are covered in Article 430. Section 210.2 provides a cross-reference table to steer the Code user to the correct article for *specific purpose branch circuits.*

FIGURE 9–4 The conductors from the final overcurrent device, fuse, or circuit breaker to the receptacles, lighting outlets, hardwired equipment and all other outlets are branch circuit conductors.

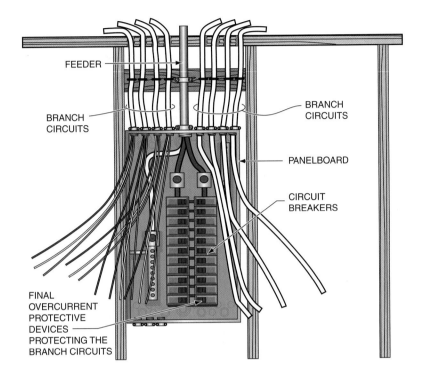

Article 210 is subdivided into three logical parts as follows:

Part I General Provisions

- Ratings
- Identification
- Voltage Limitations
- Receptacle Requirements
- GFCI Requirements
- AFCI Requirements
- Required Branch Circuits

Part II Branch Circuit Ratings

- Conductor Ampacity and Size
- Overcurrent Protection
- Outlet Devices
- Permissible Loads
- Common Area Branch Circuits

Part III Required Outlets

- Dwelling Units
- Guest Rooms
- Show Windows
- Lighting Required
- HVAC Outlet Required

DidYouKnow?

Article 100 defines the term "branch circuit." In addition to this basic definition, Article 100 provides four definitions for specific types of branch circuits.
- Branch circuit, appliance
- Branch circuit, general-Purpose
- Branch circuit, individual
- Branch circuit, multiwire

DidYouKnow?

In order to determine the minimum ampacity (size) of service, feeder, and branch circuit conductors, the requirements of Article 220 must be applied along with requirements in Articles 210, 215, 225, and 230.

Article 215 Feeders

The definition of the term *feeder* exists in Article 100 and is defined as *all circuit conductors between the service equipment, the source of a separately derived system, or other power supply source and the final branch-circuit overcurrent device* (Figure 9–5).

The term *feeder* is often misused as a "sub-feeder." This term does not exist in the *NEC*®. The three primary types of conductors addressed in the *NEC*® are service, feeder, and branch circuit. Service conductors may only originate at a utility-owned and-supplied source, and end where disconnecting means and overcurrent protection are provided. Branch circuits begin at the final overcurrent protective device and end at the outlet or utilization equipment supplied. Feeders are, very simply, all of the conductors in between the service equipment and the final overcurrent protective device. A hierarchy does not exist for feeders, therefore, there is no such thing as a "sub-feeder."

Article 215 provides requirements for the following:

(1) Installation requirements for feeders

(2) Overcurrent protection requirements for feeders

(3) Minimum size and ampacity requirements for feeders

Article 220 Branch Circuit, Feeder and Service Calculations

Article 220 is subdivided into five logical parts as follows:

Part I General

Part II Branch Circuit Load Calculations

Part III Feeder and Service Load Calculations

FIGURE 9–5   All conductors between the final overcurrent protective device and the service, or other power supply, are feeder conductors.

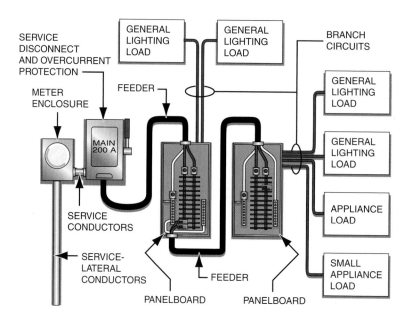

Part IV Optional Feeder and Service Load Calculations

Part V Farm Load Calculation

Article 220 provides requirements for the following:

(1) Computing branch circuit loads

(2) Computing feeder loads

(3) Computing service loads

Article 225 Outside Branch Circuits and Feeders

Articles 210 and 215 provide the basic requirements for *branch circuits* and *feeders*. This article provides specific requirements for *outside branch circuits* and *feeders* (Figure 9–6). Article 225 provides specific requirements for outside branch circuits and feeders as follows:

(1) Outside branch circuits and feeders run on or between buildings

(2) Outside branch circuits and feeders on structures

(3) Outside branch circuits and feeders on poles on the premises

(4) Outside branch circuits and feeders supplying electric equipment and wiring for the supply of utilization equipment that is located on or attached to the outside of buildings, structures, or poles

DidYouKnow?

In addition to the requirements of Articles 210 and 215, branch circuits and feeders run outside of a building or structure must also comply with Article 225.

FIGURE 9–6   Article 225 applies to all feeders and branch circuits located outdoors.

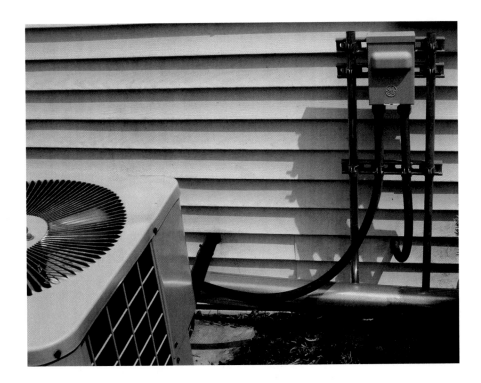

Section 225.2 provides a cross-reference table to steer the Code user to the correct article where the *NEC*® provides additional requirements for outside branch circuits and feeders.

Article 225 is subdivided into three logical parts as follows:

Part I General

This part provides general Information for Outside Branch Circuits and Feeders, including:

- Conductor Size and Support
- Outdoor Lighting Equipment
- Overcurrent Protection
- Wiring on Buildings
- Entrance/Exit of Circuits
- Open Conductor Spacing/Support
- Support Over Buildings
- Point of Attachment
- Means of Attachment
- Clearance from Ground
- Protection of Conductors
- Raceways/Cables on Exterior
- Vegetation as Support

Part II More than One Building or Other Structure

This part addresses outside branch circuits and feeders that supply a separate building or structure. Note that a building or structure that is supplied by an outdoor branch circuit or feeder must comply with rules very similar to those for a service-supplied building or structure.

This part provides requirements for outside branch circuits and feeders that supply more than one building or structure. These basic requirements include the following:

- Number of Supplies
- Location of Disconnecting Means
- Maximum Number of Disconnects
- Grouping of Disconnects
- Accessibility of Disconnects
- Identification of Disconnects
- Disconnect Rating
- Access to Overcurrent Protective Devices

Part III Over 600 Volts

Outside branch circuits and feeders operating at over 600 volts are addressed in Part III of Article 225. This part provides requirements very similar to those in Parts I and II, modified for systems operating at over 600-volts.

Article 230 Services

Proper application of Article 230 requires that the Code user is familiar with several Article 100 definitions related to services (Figure 9–7). The definition of the term *service* exists in Article 100 and is defined as *the conductors and equipment for delivering electric energy from the serving utility to the wiring system of the premises served.* The definition of *service* includes all conductors from the service point to the service disconnect and overcurrent protection. This is seen in the definition of *service conductors* which is defined as *the conductors from the service point to the service disconnecting means.*

FIGURE 9-7

Article 230 covers all conductors and equipment from the service point to the service disconnect and overcurrent protection.

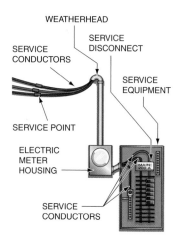

WEATHERHEAD

SERVICE DISCONNECT

SERVICE CONDUCTORS

SERVICE EQUIPMENT

SERVICE POINT

ELECTRIC METER HOUSING

SERVICE CONDUCTORS

This definition would include *service drop (overhead)* and *service lateral (underground)* conductors. The requirements of Article 230 have been specifically developed to protect persons and property from service conductors that are considered to be unprotected and, in most cases, can only be disconnected by the serving utility.

Article 230 is subdivided into eight logical parts, with each part logically named for the requirements contained as follows:

Part I General
- Number of Services
- Service Conductors Considered as Outside
- Raceway Seals
- Clearance of Service Conductors from Building Openings
- Vegetation as Support

Part II Overhead Service Drop Conductors
- Insulation or Covering
- Conductor Size and Rating
- Clearance Above Ground and Rooftops
- Point of Attachment to Building
- Means of Attachment
- Service Masts as Supports
- Supports over a Building

Part III Underground Service Lateral Conductors
- Insulation
- Conductor Size and Rating
- Splices
- Protection

Part IV Service Entrance Conductors
- Number of Sets Served
- Insulation
- Protection
- Conductor Size and Rating
- Permitted Wiring Methods
- Splices
- Cable Trays
- Supports
- Raceways Arranged to Drain
- Overhead Connections Service Head/Gooseneck, Drip-loops
- 4-wire Delta Configured Systems

Part V Service Equipment,–General
- Enclosure Requirements
- Marking of Equipment, Suitable as "Service Equipment"

Part VI Service Equipment,–Disconnecting Means
- Disconnect Requirement
- Location of Disconnect

Did You Know?

Article 230 is logically separated into eight parts which are titled to illustrate the information and requirements they contain. This is an excellent example of how the structure or the *NEC®* provides the Code user with a roadmap to locate information and requirements.

Did You Know?

Chapter 2 of the *NEC*® is titled "Wiring and Protection."

- Marking
- Accessibility of Disconnect
- Maximum Number of Disconnects
- Grouping of Disconnects
- Operation of Disconnects
- Indicating (Identifies Open or Closed)
- Rating of Disconnect (amps)
- Equipment Permitted on the Line Side (Upstream) of the Service Disconnect

Part VII Service Equipment, Overcurrent Protection

- Where Required
- Location
- Locking of Service OCPD
- Ground Fault Protection of Equipment (GFPE)

Part VIII Services Exceeding 600 Volts, Nominal

- General Requirements
- Isolating Switches
- Disconnects
- Protection
- Surge Arrestors
- Metal Enclosed Switchgear
- Services over 35,000 Volts

Did You Know?

Chapter 2 requirements for "Wiring" are located in Articles 200, 210, 215, 220, 225, and 230.

Did You Know?

Chapter 2 requirements for "Protection" are located in Articles 240, 250, 280, and 285.

PROTECTION

In the planning stages of any electrical installation, protection must be part of the design incorporated into the electrical drawings. These drawings will provide the installer with the necessary information to BUILD an installation that is properly protected. All electrical installations must provide overcurrent protection for all current-carrying conductors. All current-carrying conductors must be protected in accordance with the conductor ampacity and conditions of use in accordance with Article 240 Overcurrent Protection.

Grounding of electrical systems is designed to limit the voltage imposed by lighting, line surges, or unintentional contact with higher voltage lines and to stabilize the voltage to earth during normal operation. Grounding of equipment is provided to limit the voltage to ground on all equipment. Systems and circuits are required to be grounded in accordance with Article 250 Grounding and Bonding.

Protection from surge voltages is provided in electrical systems through the use of Article 280 Surge Arrestors, Over 1 kV. Protection from voltage surges at levels closer to nominal voltage is provided for by the application of Article 285, Surge-Protective Devices (SPDs), 1 kV or Less.

Code-compliant electrical installations must provide adequate protection of persons and property in accordance with the *NEC*®. The required protection for electrical installations that must be planned include the following keywords/clues:

- Overcurrent protection, fuses, circuit breakers
- Grounding and bonding of systems and circuits

- Grounding and bonding raceways, equipment, or enclosures
- Surge arrestors, over 1 kV
- Surge-protective devices (SPDs), 1 kV or less

Articles dedicated to protection are shown in Table 9–2:

TABLE 9–2	Layout of NEC®, Chapter 2: Wiring and Protection
NEC® Title: Codeology Title: Chapter Scope:	Wiring and Protection Plan Information and Rules on Wiring and Protection of Electrical Installations
	Protection
Article	Article Title
240	Overcurrent Protection
250	Grounding and Bonding
280	Surge Arrestors, Over 1 kV
285	Surge-Protective Devices (SPDs), 1 kV or Less

Article 240 Overcurrent Protection

The definition of the term *overcurrent* exists in Article 100 and is defined as *any current in excess of the rated current of equipment or the ampacity of a conductor. It may result from overload, short circuit, or ground fault.*

This basic definition must be understood to properly apply the provisions of Article 240. All conductors and equipment will have ratings for the amount of current (amps) that the conductor or equipment can handle without suffering damage. An overcurrent occurs when one of three incidents happen: an (1) overload, (2) short circuit, or (3) ground fault.

Overloads occur when a conductor or equipment are subjected to a current above their rating, which if allowed to continue for too long would cause damage or dangerous overheating. A fault, such as a short circuit or ground fault, is not an overload. A conductor rated at 20 amps with 22 amps of current flowing would be experiencing an overload. An overload never leaves the circuit path. Current continues to flow from the source through the circuit conductors and load back to the source.

A short circuit is when the current leaves the normal circuit path and takes a short cut. Short circuits occur when current-carrying conductors make contact with each other, which creates a shortcut for current flow. Any combination of two or more circuit conductors (current-carrying) in contact is a short circuit. Current-carrying conductors include all ungrounded (hot) and grounded (neutral, in most cases) conductors.

A ground fault is a form of a short circuit. The current takes a shortcut on a grounded component, such as a raceway, enclosure, building steel, or an equipment grounding or bonding conductor.

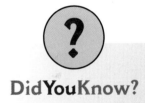

DidYouKnow?

Article 240 requires that, in general, all ungrounded conductors be provided with overcurrent protection at the point where the conductors receive their supply.

The scope of Article 240 Overcurrent Protection is to provide general requirements for overcurrent protection and for overcurrent protective devices. Overcurrent protective devices (OCPD) include but are not limited to fuses and circuit breakers (Figures 9–8a and b). These devices are designed to provide protection from overloads, short circuits, and ground faults. Article 240 is subdivided into nine logical parts as follows:

Part I General

- Definitions
- Cross-reference Other Articles
- Protection of Conductors
- Protection of Flexible Cords/Cables
- Standard OCPD amp Ratings
- Fuses or Circuit Breakers in Parallel
- Supplementary Overcurrent Protection
- Thermal Devices
- Electrical System Coordination
- Ground Fault Protection of Equipment

FIGURE 9–8 Fuses and circuit breakers provide overcurrent protection for conductors and equipment.

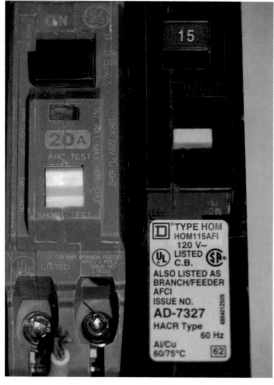

(a)

(b)

Part II Location

- Ungrounded Conductors
- Location of Overcurrent Protection in Circuit, Tap Rules
- Grounded Conductors
- Location/Accessibility of OCPDs

Part III Enclosures

- General Protection and Operation
- Damp or Wet Locations
- Vertical Position

Part IV Disconnecting and Guarding

- Disconnects for Fuses
- Arcing or Suddenly Moving Parts

Part V Plug Fuses, Fuseholders, and Adapters

- General Application
- Edison Base Fuses
- Type "S" Fuses

Part VI Cartridge Fuses and Fuseholders

- General Application
- Classification

Part VII Circuit Breakers

- Method of Operation
- Indicating
- Non-Tamperable
- Marking
- Applications
- Series Ratings

Part VIII Supervised Industrial Installations

This part addresses only those portions of a building or structure that meet the conditions of the definition of *supervised industrial installation* defined in 240.2.

Part IX Overcurrent Protection Over 600 Volts, Nominal

This part is limited only to feeders and branch circuits operating at over 600 volts nominal.

Article 250 Grounding and Bonding

The scope of Article 250 is provided in the first section of the article, 250.1 Scope. *This article covers general requirements for grounding and bonding of electrical installations* (Figures 9–9a and b), *and the specific requirements in (1) through (6).*

(1) *Systems, circuits, and equipment required, permitted, or not permitted to be grounded*

(2) *Circuit conductor to be grounded on grounded systems*

(3) *Location of grounding connections*

FIGURE 9–9

Article 250 covers protection requirements through provisions for grounding and bonding of electrical systems and installations.

(a)

(b)

(4) Types and sizes of grounding and bonding conductors and electrodes

(5) Methods of grounding and bonding

(6) Conditions under which guards, isolation, or insulation may be substituted for grounding

The *National Electrical Code*® is a "prescriptive" installation document. This means that the *NEC*® will give specific requirements for compliance to the installer without stating the reason for compliance. The general rule throughout the *NEC*® is not to explain individual requirements for two reasons. The first reason is that, as per section 90.1(C), the *NEC*® is not intended to be a design manual or an instruction manual for untrained persons. The second reason is due to the size of the *NEC*®; if all requirements were explained, the *NEC*® would be five times as thick in size. However, in Article 250 Grounding and Bonding, Section 250.4 explains to the user of the *NEC*® exactly what grounding and bonding of electrical systems are required to accomplish.

The following outline of 250.4 illustrates what is explained in this section:

250.4 General Requirements for Grounding and Bonding

(A)   Grounded Systems

   (1)   Electrical System Grounding

   (2)   Grounding of Electrical Equipment

   (3)   Bonding of Electrical Equipment

   (4)   Bonding of Electrically Conductive Materials and Other Equipment

   (5)   Effective Ground-Fault Current Path

(B)   Ungrounded Systems

   (1)   Grounding Electrical Equipment

   (2)   Bonding of Electrical Equipment

   (3)   Bonding of Electrically Conductive Materials and Other Equipment

   (4)   Path for Fault Current

Article 250 is subdivided into ten logical parts as follows:

DidYouKnow?

Article 250 contains a substantial amount of information and requirements for grounding and bonding, spanning 31 pages in the 2008 *NEC*®. It is essential for the Code user to understand the structure of the *NEC*® and access the right part of Article 250 before attempting to locate needed information or requirements.

Part I General

- Definitions
- Cross-reference Other Articles
- Reasons for Grounding and Bonding
- Connections
- Protection of Clamps/Fittings
- Clean Surfaces

Part II System Grounding

- Systems Requiring Grounding
- Systems *NOT* Requiring Grounding
- Circuits not Permitted to Be Grounded
- Grounding AC Services
- Conductor to be Grounded
- Main Bonding Jumpers

- Separately Derived Systems
- Two or More Buildings with Common Service
- Portable Generators
- High-Impedance Systems

Part III Grounding Electrode System and Grounding Electrode Conductor

- Outline of the Grounding Electrode System
- Permitted Electrodes
- Installation of Grounding Electrode System
- Common Electrodes
- Supplementary Electrodes
- Resistance of Electrodes
- Air Terminals
- Size of Grounding Electrode Conductor
- Connection of Grounding Electrode Conductors

Part IV Enclosure, Raceway, and Service Cable Connections

- Service Raceways and Enclosures
- Underground Service Cable and Conduit
- Other Enclosures and Raceways

Part V Bonding

- Services
- Other Systems
- Other Enclosures
- Over 250 Volts
- Loosely Jointed Raceways
- Hazardous Locations
- Equipment Bonding Jumpers, Supply and Load Side
- Piping Systems and Exposed Structural Steel
- Lightning Protection Systems

Part VI Equipment Grounding and Equipment Grounding Conductors

- Equipment Fastened in Place
- Cord-and-Plug-Connected Equipment
- Nonelectric Equipment
- Types of Equipment Grounding Conductors
- Identification of Equipment Grounding Conductors and Device Terminals
- Installation
- Size of Equipment Grounding Conductors

Part VII Methods of Equipment Grounding

- Connections
- Short Sections of Raceway
- Ranges/Clothes Dryer Frames
- Equipment Fastened in Place
- Cord-and-Plug-Connected Equipment
- Use of Grounded Conductor

DidYouKnow?

When the user of the *NEC*® has a question on how to size an equipment grounding conductor, the answer will be located in Part VI of Article 250.

- Receptacle Grounding
- Attachment to Boxes

Part VIII Direct Current Systems
- Circuits and Systems to Be Grounded
- Point of Connection
- Size of Grounding Electrode Conductor
- Bonding Jumpers

Part IX Instruments, Meters, and Relays
- Transformer Circuits and Cases
- Cases of Equipment at Over 1,000 Volts
- Grounding Conductors

Part X Grounding of Systems and Circuits of 1 kV and Over (High Voltage)
- General Requirements
- Derived Neutral Systems
- Grounding
- Solidly Grounded Neutral Systems
- Impedance Grounded Neutral Systems
- Portable or Mobile Equipment

Article 280 Surge Arrestors, Over 1 kV

The definition of *surge arrestors* exists in Section Article 100 and is defined as *a protective device for limiting surge voltages by discharging or bypassing surge current; it also prevents continued flow of follow current while remaining capable of repeating these functions* (Figure 9–10).

Article 280 is subdivided into three logical parts as follows:

Part I General
- Uses Not Permitted
- Listing
- Number of Surge Arrestors Required
- Selection

Part II Installation
- Location of Surge Arrestors
- Routing Surge Arrestor Connections

Part III Connecting Surge Arrestors
- Services of Over 1,000 Volts
- Load Side Installation of Over 1,000 Volts
- Circuits 1,000 Volts and Over
- Grounding

Article 285 Surge-Protective Devices (SPDs), 1 kV or Less

The definition of *surge-protective devices (SPDs)* exists in Article 100 and is defined as *a protective device for limiting transient voltages by diverting or limiting surge current; it also prevents continued flow of follow current*

DidYouKnow?

Lightning is one of the most obvious sources of a power surge but is not the only source. The requirements of Article 280 are limited to applications over 1,000 volts.

FIGURE 9-10 Surge arrestors limit surge voltages by discharging or bypassing surge currents.

while remaining capable of repeating these functions and is designated as follows:

Type 1: Permanently connected SPDs intended for installation between the secondary of the service transformer and the line side of the service disconnect overcurrent device.

Type 2: Permanently connected SPDs intended for installation on the load side of the service disconnect overcurrent device, including SPDs located at the branch panel.

Type 3: Point of utilization SPDs.

Type 4: Component SPDs, including discreet components, as well as assemblies.

Article 285 is subdivided into three logical parts as follows:

Part I General

- Listing
- Uses Not Permitted
- Number Required
- Listing Requirements
- Short Circuit Ratings

Part II Installation

- Location
- Routing of Connections

FIGURE 9-11

Surge-protective devices (SPDs) provide protection from an overvoltage at levels much closer to the operating voltage than surge arrestors.

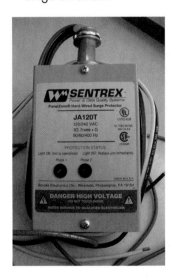

Part III Connecting SPDs
- Connection of SPDs
- Grounding

KEY WORDS/CLUES FOR CHAPTER 2, "PLAN"

A list of key words and clues for Chapter 2 is shown in Table 9–3:

TABLE 9–3	Clues and Key Words	
Wiring	**Protection**
Branch circuits, all types GFCI	Overcurrent protection, fuses, circuit breakers
Branch circuits, indoor/outdoor	Grounding systems and circuits
Required outlets AFCI	Grounding and bonding raceways, equipment or enclosures
Feeders, indoor/outdoor	Surge arrestors
Service/s service equipment	Surge-protective devices
Calculations, computed load	Equipment grounding conductors
Grounded conductors, neutral	Location of overcurrent protective devices

THINK "PLAN" AND GO TO CHAPTER 2

Refer to Chapter 2 when your question or need within the *NEC*® deals with:
- Wiring, *NEC*® Current-Carrying Conductor Names, Branch Circuit, Feeder, Service or Tap Conductor
- Grounded Conductors
- Ground Fault Circuit Interrupters (GFCI)
- Arc Fault Circuit Interrupters (AFCI)
- Required Outlets
- Calculations, Computed Loads
- Outside Branch Circuits and Feeders
- Protection, Overcurrent, Grounding, Surge Arrestors, and Surge-Protective Devices
- Fuses and Circuit Breakers
- Grounding and Bonding Conductors
- Grounding Electrodes

SUMMARY

Chapter 2, in accordance with Section 90.3, applies generally to all electrical installations.

Using the Codeology method, this chapter is given the nickname of the "Plan" chapter, due to the scope covered. The *NEC®* title for Chapter 2 is "Wiring and Protection." From this title the scope of this chapter is described as "Information and Rules on *Wiring and Protection* of Electrical Installations."

Chapter 2 covers the entire electrical system from the service point (connection to the utility) to the last receptacle or other outlet in the electrical system.

This chapter, in accordance with its scope of "Wiring," provides detailed requirements for all current-carrying conductors. The *NEC®* names and defines these conductors to provide clarity. These *NEC®* current-carrying conductor names include service, feeder, branch circuit, and tap conductors.

Chapter 2, in accordance with its scope of "Protection," provides detailed requirements to protect the entire electrical system. These protection requirements include overcurrent, grounding, bonding, surge arrestors, and surge-protective devices.

See Figure 9–12 for an example of how Chapter 2 along with Chapters 1, 3 and 4 build the foundation or backbone of all electrical installations.

REVIEW QUESTIONS

1. Does Chapter 2 of the *NEC®* apply to all electrical installations covered by the *NEC®*?

2. Chapter 2 is subdivided into how many articles?

3. The scope of Chapter 2 of the *NEC®* is separated into two areas with six articles covering _____ and four articles covering _____.

4. What part of which article in Chapter 2 would address an outdoor branch circuit serving a second structure?

5. Chapter 2 of the *NEC®* addresses wiring, which would include calculating the size of conductors. What part of which article would apply if one were calculating loads for a single-family dwelling unit service?

6. Does 210.8(B) address ground fault circuit interrupter requirements in dwelling units?

7. The location of Section 250.166 limits the application of this section to what type of system?

8. How many parts is Article 240 subdivided into?

9. Does Part V of Article 240 apply to cartridge-type fuses?

10. Which part of Article 250 would contain requirements for grounding electrode conductors?

FIGURE 9–12 Chapter 2 will apply generally in all electrical installations.

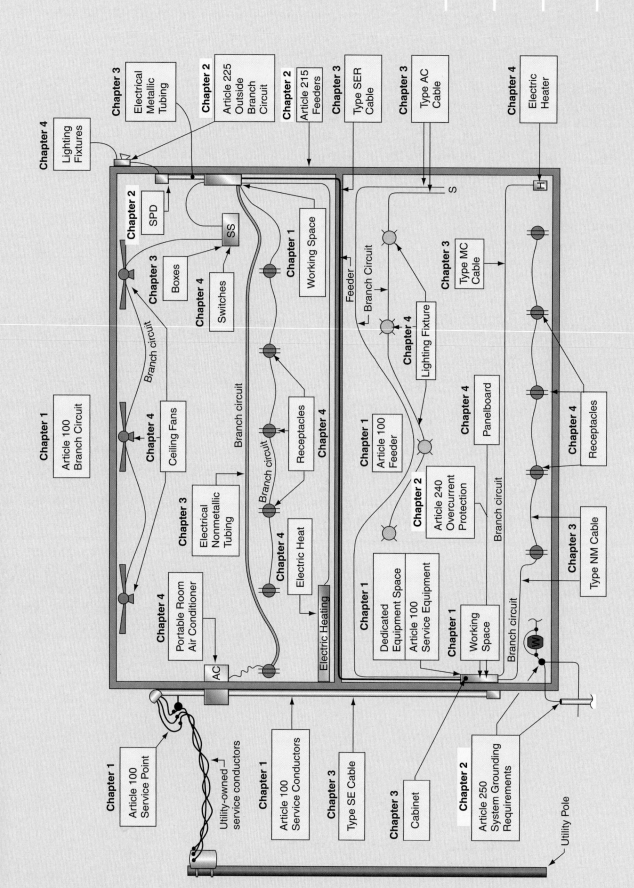

Chapter 4 — Lighting Fixtures

Chapter 3 — Electrical Metallic Tubing

Chapter 2 — Article 225 Outside Branch Circuit

Chapter 2 — Article 215 Feeders

Chapter 3 — Type SER Cable

Chapter 3 — Type AC Cable

Chapter 4 — Electric Heater

Chapter 2 — SPD

Chapter 3 — Boxes

Chapter 4 — Switches

Chapter 1 — Working Space

Chapter 3 — Type MC Cable

Chapter 1 — Article 100 Branch Circuit

Chapter 4 — Ceiling Fans

Chapter 3 — Electrical Nonmetallic Tubing

Chapter 4 — Receptacles

Chapter 4 — Lighting Fixture

Chapter 4 — Panelboard

Chapter 1 — Article 100 Feeder

Chapter 4 — Portable Room Air Conditioner

Chapter 4 — Electric Heat

Chapter 2 — Article 240 Overcurrent Protection

Chapter 4 — Receptacles

Chapter 1 — Dedicated Equipment Space, Article 100 Service Equipment

Chapter 1 — Working Space

Chapter 3 — Type NM Cable

Chapter 1 — Article 100 Service Point

Utility-owned service conductors

Chapter 1 — Article 100 Service Conductors

Chapter 3 — Type SE Cable

Chapter 3 — Cabinet

Chapter 2 — Article 250 System Grounding Requirements

Utility Pole

PRACTICE PROBLEMS

The following questions and steps to find the answer are designed to illustrate the Codeology method. While the steps to find your answer may seem lengthy, they are designed to illustrate the thought process to find the answer. The easiest way to make the Codeology method work is to silently "talk to yourself." Walk through these steps by silently talking to yourself and Codeology will be a natural response for quickly and accurately finding needed information in the *NEC*®.

Read the following questions and follow step by step using your codebook and the Codeology method.

1. When calculating the lighting load for dwelling unit branch circuits, _____ volt-amps per square foot is required.

Step #1

Qualify the question or need, look for key words
   This question is about calculating a load.
   Key words: "*calculating the lighting load*"
   Think wiring, think PLAN

Step #2

Go to the table of contents, Chapter 2, the Plan chapter
   Find the correct article   **220 Calculations**

Step #3

Further qualify your question or need and get in the right part of the article
   Key words: "*volt amps per square foot is general calculations for branch circuit loads.*"

   Look in   **Part II Branch Circuit Load Calculations**

Step #4

Read each section title until the correct section is found
   Identify correct section or subdivision and find answer
   Correct section/subdivision
   Section   **220.12 Lighting Load for Specified Occupancies**
   Table   **Table 220.12 General Lighting Loads by Occupancy**

2. Each feeder disconnect rated _____ amps or more installed in a solidly grounded wye-connected system of more than 150 volts to ground but not more than 600 volts phase to phase shall be provided with ground fault protection of equipment.

Step #1

Qualify the question or need, look for key words
   This question is about a feeder.
   Key words: "*disconnect rating, ground fault protection of equipment*"
   Think wiring, think PLAN

Step #2

Go to the table of contents, Chapter 2, the Plan chapter
   Find the correct article   **215 Feeders**

Step #3

Further qualify your question or need and get in the right part of the article
   Key words: "*ground fault protection of equipment*"

   Article 215 is not subdivided into parts.

Step #4

Read each section title until the correct section is found
   Identify correct section or subdivision and find answer
   Correct section/subdivision
   Section   **215.10 Ground Fault Protection of Equipment**

3. In general, interior metal water piping located more than _____ feet from the point of entrance to the building shall not be used as a part of the grounding electrode system.

Step #1

Qualify the question or need, look for keywords
   This question is about grounding.
   Key words: "*grounding electrode system*" and "*interior metal water piping*"
   Think protection, think PLAN

Step #2

Go to the table of contents, Chapter 2, the Plan chapter

Find the correct article   **250 Grounding and Bonding**

Step #3

Further qualify your question or need and get in the right part of the article

Key words: *"grounding electrode system"* and "interior *metal water piping"*

Look in   **Part III Grounding Electrode System and Grounding Electrode Conductor**

Step #4

Read each section title until the correct section is found

Identify correct section or subdivision and find answer

Correct section/subdivision

Section   **250.52 Grounding Electrodes**

First-Level Subdivision   **(A) Electrodes Permitted for Grounding**

Second-Level Subdivision   **(1) Metal Underground Water Pipe**

---

4. The service disconnecting means serving a one-family dwelling unit shall have an ampere rating of not less than _____ amps.

Step #1

Qualify the question or need, look for key words

This question is about a service.

Key words: *"service disconnecting means"* and *"ampere rating"*

Think wiring, think PLAN

Step #2

Go to the table of contents, Chapter 2, the Plan chapter

Find the correct article   **230 Services**

Step #3

Further qualify your question or need and get in the right part of the article

Key words: *"disconnecting means," "ampere rating"*, and *"one-family dwelling"*

Look in   **Part VI Service Equipment— Disconnecting Means**

Step #4

Read each section title until the correct section is found

Identify correct section or subdivision and find answer

Correct section/subdivision

Section   **230.79 Rating of Service Disconnecting Means**

First-Level Subdivision   **(C) One-Family Dwelling**

---

5. A surge-protective device shall not be permitted for use on circuits exceeding _____ volts.

Step #1

Qualify the question or need, look for key words

This question is about a transient voltage surge suppresor.

Key words: *"surge-protective device"*

Think protection, think PLAN

Step #2

Go to the table of contents, Chapter 2, the Plan chapter

Find the correct article   **285 Surge-Protective Devices (SPDs), 1 kV or Less**

Step #3

Further qualify your question or need and get in the right part of the article

Key words: *"shall not be permitted"* and *"circuits exceeding _____ volts"*

Look in   **Part I General, uses permitted and not permitted is a general question**

Step #4

Read each section title until the correct section is found

Identify correct section or subdivision and find answer

Correct section/subdivision

Section   **285.3 Uses Not Permitted**

6. A 125-volt, single-phase, 15- or 20-ampere-rated receptacle outlet is required to be installed at an accessible location for the servicing of heating, air-conditioning, and refrigeration equipment. The receptacle shall be located on the same level and within _____ feet of the equipment.

Step #1

Qualify the question or need, look for key words
This question is about a "required outlet"; outlets are supplied by branch circuits.
Key words: *"receptacle outlet is required"*
Think wiring, think PLAN

Step #2

Go to the table of contents, Chapter 2, the Plan chapter
Find the correct article   **210 Branch Circuits**

Step #3

Further qualify your question or need and get in the right part of the article
Key words: *"receptacle outlet is required"*, *"heating, air-conditioning"*, and *refrigeration equipment"*
Look in   **Part III Required Outlets**

Step #4

Read each section title until the correct section is found
Identify correct section or subdivision and find answer
Correct section/subdivision
Section   **210.63 Heating, Air-Conditioning, and Refrigeration Equipment Outlet**

7. An insulated grounded conductor of 6 AWG or smaller shall be identified by a continuous white or gray outer finish or by _____ continuous white stripes on other than green insulation along its entire length.

Step #1

Qualify the question or need, look for key words
This question is about a grounded conductor.
Key words: *"grounded conductor"* and *"shall be identified"*
Think wiring, think PLAN

Step #2

Go to the table of contents, Chapter 2, the Plan chapter
Find the correct article   **200 Use and Identification of Grounded Conductors**

Step #3

Further qualify your question or need and get in the right part of the article
Key words: *"grounded conductor"* and *"shall be identified"*
Article 200 is not subdivided into parts.

Step #4

Read each section title until the correct section is found
Identify correct section or subdivision and find answer
Correct section/subdivision
Section   **200.6 Means of Identifying Grounded Conductors**
First-Level Subdivision   **(A) Sizes 6 AWG or Smaller**

8. Cartridge fuses shall be plainly marked, either by printing on the fuse barrel or by a label attached to the barrel showing ampere rating, voltage rating, interrupting rating when other than _____ amperes, current limiting when applicable, and the name or trademark of the manufacturer.

Step #1

Qualify the question or need, look for key words
This question is about fuses that are overcurrent devices.
Key words: *"cartridge fuses"* and *"shall be plainly marked"*
Think protection, think PLAN

Step #2

Go to the table of contents, Chapter 2, the Plan chapter
Find the correct article   **240 Overcurrent Protection**

Step #3

Further qualify your question or need and get in the right part of the article

Key words: "cartridge *fuses*" and *"shall be plainly marked"*

Look in   **Part VI Cartridge Fuses and Fuse-holders**

Step #4

Read each section title until the correct section is found

Identify correct section or subdivision and find answer

Correct section/subdivision

Section   **240.60 General**

First-Level Subdivision   **(C) Marking**

---

9. Conductors used to connect a surge arrester in a 15kV system to line or bus and to ground shall not be any longer than necessary and shall avoid unnecessary _____.

Step #1

Qualify the question or need, look for key words

This question is about a surge arrestor.

Key words: *"Conductors used to connect a surge arrester"*

Think protection, think PLAN

Step #2

Go to the table of contents, Chapter 2, the Plan chapter

Find the correct article   **280 Surge Arrestors, Over 1 kV**

Step #3

Further qualify your question or need and get in the right part of the article

Key words: *"Conductors used to connect a surge arrester"* and *"to ground"*

Look in   **Part II Installation; connecting conductors infers installation**

Step #4

Read each section title until the correct section is found

Identify correct section or subdivision and find answer

Correct section/subdivision

Section   **280.12 Routing of Surge Arrestor Grounding Conductors**

---

10. Branch circuits installed in raceways on exterior surfaces of buildings or other structures shall be raintight and arranged to _____.

Step #1

Qualify the question or need, look for key words

This question is about a branch circuit installed outside of a building.

Key words: *"branch circuits"* and *"in raceways on exterior surfaces of buildings"*

Think wiring, think PLAN

Step #2

Go to the table of contents, Chapter 2, the Plan chapter

Find the correct article   **225 Outside Branch Circuits and Feeders**

Step #3

Further qualify your question or need and get in the right part of the article

Key words: *"in raceways on exterior surfaces of buildings"*

Look in   **Part I General**

Step #4

Read each section title until the correct section is found

Identify correct section or subdivision and find answer

Correct section/subdivision

Section   **225.22 Raceways on Exterior Surfaces of Buildings or Other Structures**

---

11. When calculating the size of a dwelling unit service, a load of not less than _____ volt-amperes shall be included for each 2-wire laundry branch circuit installed.

Step #1

Qualify the question or need, look for key words

This question is about a calculation

Key words: *"calculating the size"* and *"dwelling unit service"*

Think wiring, think PLAN

Go to the table of contents, Chapter 2, the Plan chapter

Find the correct article   **220 Branch Circuit, Feeder and Service Calculations**

Further qualify your question or need and get in the right part of the article

Key words: *"dwelling unit service"* and *"for each 2-wire laundry branch circuit"*

Look in   **Part III Feeder and Service Load Calculations**

Read each section title until the correct section is found

Identify correct section or subdivision and find answer

Correct section/subdivision

Section   **220.52 Small Appliance and Laundry Loads—Dwelling Unit**

First-Level Subdivision   **(B) Laundry Circuit Load**

---

12. In dwelling units and guest rooms of hotels, motels, and similar occupancies, the branch circuit voltage shall not exceed _____ volts, nominal, between conductors that supply lighting fixtures.

Qualify the question or need, look for key words

This question is about a branch circuit.

Key words: *"branch circuit voltage shall not exceed"*

Think wiring, think PLAN

Go to the table of contents, Chapter 2, the Plan chapter

Find the correct article   **210 Branch Circuits**

Further qualify your question or need and get in the right part of the article

Key words: *"voltage* shall *not exceed"*

Look in   **Part I General Provisions**

Read each section title until the correct section is found

Identify correct section or subdivision and find answer

Correct section/subdivision

Section   **210.6 Branch-Circuit Voltage Limitations**

First-Level Subdivision   **(A) Occupancy Limitation**

---

13. Circuit breakers used as switches in high-intensity discharge lighting circuits shall be listed and shall be marked as _____.

Qualify the question or need, look for key words

This question is about a circuit breaker, which is an overcurrent device.

Key words: *"Circuit breakers used as switches"* and *"shall be marked"*

Think protection, think PLAN

Go to the table of contents, Chapter 2, the Plan chapter

Find the correct article   **240 Overcurrent Protection**

Further qualify your question or need and get in the right part of the article

Key words: *"circuit breakers used as switches"* and *"shall be marked"*

Look in   **Part VII Circuit Breakers**

Read each section title until the correct section is found

Identify correct section or subdivision and find answer

Correct section/subdivision

Section   **240.83 Marking**

First-Level Subdivision   **(D) Used as Switches**

14. When a single equipment grounding conductor is run with multiple circuits in the same raceway or cable, it shall be sized for the _____ overcurrent device protecting conductors in the raceway or cable.

Step #1

Qualify the question or need, look for key words
This question is about grounding.

Key words: "single equipment grounding conductor"

Think protection, think PLAN

Step #2

Go to the table of contents, Chapter 2, the Plan chapter

Find the correct article   **250 Grounding and Bonding**

Step #3

Further qualify your question or need and get in the right part of the article

Key words: "single equipment grounding conductor," with "multiple circuits in the same raceway or cable," and "shall be sized"

Look in   **Part VI Equipment Grounding and Equipment Grounding Conductors**

Step #4

Read each section title until the correct section is found

Identify correct section or subdivision and find answer

Correct section/subdivision

Section   **250.122 Size of Equipment Grounding Conductors**

First-Level Subdivision   **(C) Multiple Circuits**

15. When an isolating switch is installed in part of the service equipment and the nominal voltage exceeds 600-volts, the isolating switch shall be accessible to _____ _____ only.

Step #1

Qualify the question or need, look for key words
This question is about a service.

Key words: "service equipment," "exceeds 600-volts" and "isolating switch"

Think wiring, think PLAN

Step #2

Go to the table of contents, Chapter 2, the Plan chapter

Find the correct article   **230 Services**

Step #3

Further qualify your question or need and get in the right part of the article

Key words: "exceeds 600-volts" and "isolating switch"

Look in   **Part VIII Services Exceeding 600 Volts, Nominal**

Step #4

Read each section title until the correct section is found

Identify correct section or subdivision and find answer

Correct section/subdivision

Section   **230.204 Isolating Switches**

First-Level Subdivision   **(C) Accessible to Qualified Persons Only**

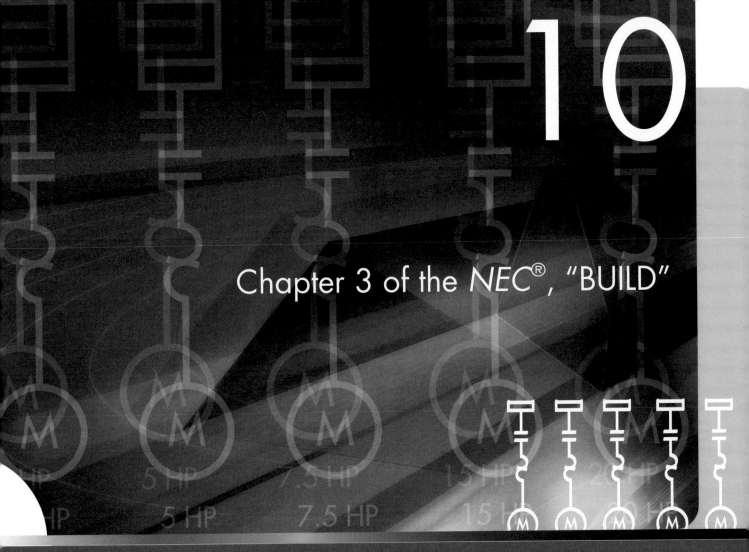

Chapter 3 of the NEC®, "BUILD"

10

OUTLINE

OBJECTIVES

After completing this unit, you should be able to:

1. Associate the Codeology title for *NEC®* Chapter 3 as "Build"
2. Understand the building and hands-on type of information and requirements for wiring methods and wiring materials contained in Chapter 3
3. Recognize key words and clues for locating answers in *NEC®* Chapter 3, the Build Chapter
4. Identify exact sections, subdivisions, list items, etc. to justify answers for all questions referring to Chapter 3 articles
5. Recognize that Chapter 3 numbering is the 300-series
6. Recognize, recall, and become familiar with articles contained in Chapter 3

OVERVIEW

The Codeology title for Chapter 3 is "Build." All electrical installations are built after they have been properly planned. Chapter 3 provides rules and information for all of the *wiring methods* and *wiring materials* used to distribute electrical energy. Electrical equipment, such as switches, receptacles, panelboards, switchboards, motors, lighting, appliances, etc., are not within the scope of this chapter. This type of equipment, which provides for control, transformation, and utilization are addressed in other chapters.

NEC®: CHAPTER 3 "BUILD"

There are forty-four articles in Chapter 3, the 300 Series. These articles address methods and materials for the distribution of electrical energy. In essence, this chapter covers all methods and materials to get electrical energy from the source to all electrical equipment and outlets in all electrical installations. Methods and materials to get electrical current from point A to point B is covered in this chapter. Building an electrical installation requires wiring methods and materials. The wiring methods and materials covered in the "Build" chapter include the following:

- Conductors
- Enclosures, cabinets, boxes
- Cable assemblies
- Circular raceways, conduits
- Other raceways
- Busways/Cablebus
- Open wiring

This chapter of the *NEC*® is dedicated to materials used to distribute electrical energy throughout buildings and structures (Figures 10–1a-c). These materials are logically broken down into two categories: wiring methods and wiring materials.

The requirements and information in this chapter are logically separated into 13 different categories as shown in Table 10–1.

FIGURE 10–1 Chapter 3 is called the Build chapter. All wiring methods and materials are located in Chapter 3.

(a)

continued

FIGURE 10–1   Chapter 3 is called the Build chapter. All wiring methods and materials are located in Chapter 3.

(b)

(c)

TABLE 10-1	Categories of *NEC®* Chapter 3

NEC® Title:	Wiring Methods and Materials
Codeology Title:	Build
Chapter Scope:	Information and Rules on *Wiring Methods and Materials* for use in Electrical Installations

Category 1	General Information for All Wiring Methods and Materials
Category 2	Conductors
Category 3	Cabinets, Boxes, Fittings, and Meter Socket/Handhole Enclosures
Category 4	Cable Assemblies
Category 5	Raceways, Circular Metal Conduit
Category 6	Raceways, Circular Nonmetallic Conduit
Category 7	Raceways, Circular Metallic Tubing
Category 8	Raceways, Circular Nonmetallic Tubing
Category 9	Factory-Assembled Power Distribution Systems
Category 10	Raceways Other than Circular
Category 11	Surface-Mounted Nonmetallic Branch Circuit Extensions
Category 12	Support Systems for Cables/Raceways
Category 13	Open-Type Wiring Methods

DidYouKnow?

Chapter 3 of the *NEC®* contains information and requirements for "wiring methods and materials" for the physical construction (building) of the electrical installation.

WIRING METHODS

Wiring methods are actually the materials used to conduct current throughout an electrical installation. Any combination of conductors and a protective means or layer to facilitate installation is a wiring method.

Cable assemblies are wiring methods (Figures 10–2a and b) For example, type AC cable consists of insulated conductors, wrapped in paper, with an outer armor of flexible metal tape and includes an internal bonding strip of copper or aluminum in intimate contact with the armor for its entire length. There are 11 types of cable assemblies in Chapter 3.

Raceways with conductors installed are also wiring methods. For example, electrical metallic tubing, type EMT with insulated type THHN conductors, installed is a wiring method. There are 22 types of raceways in Chapter 3. Raceways fall into five different categories as follows:

1. Raceways, Circular Metal Conduit (Figure 10–3)
2. Raceways, Circular Nonmetallic Conduit (Figure 10–4)
3. Raceways, Circular Metallic Tubing (Figure 10–5)
4. Raceways, Circular Nonmetallic Tubing (Figure 10–6)
5. Raceways Other than Circular (Figure 10–7)

Busways and *cablebus* are factory-assembled sections of grounded, completely enclosed, ventilated protective metal housings, containing

DidYouKnow?

Any combination of conductors and a protective means or layer to facilitate installation is a "wiring method."

FIGURE 10-2 Cable assemblies are wiring methods and are covered in Chapter 3 of the *NEC*®.

(a)

DidYouKnow?

An inspection of any electrical installation will be based upon the "wiring methods and materials" used and the occupancy or intended use.

(b)

FIGURE 10-3 Rigid Metal Conduit, Type RMC.

FIGURE 10-4 Rigid Polyvinyl Chloride Conduit, Type PVC.

FIGURE 10-5 Electrical Metallic Tubing, Type EMT.

FIGURE 10-6 Electrical Nonmetallic Tubing, Type ENT.

DidYouKnow?

An electrical installation requires (wiring methods) raceways or support methods with conductors or cable assemblies to get electrical current from point A to point B. In order to facilitate this, installation "wiring materials" such as boxes, conduit, bodies and other enclosures are used in all electrical installations.

factory-mounted, bare or insulated conductors, which are usually copper or aluminum bars, rods, or tubes (Figure 10–8).

Open wiring methods consist of concealed knob and tube wiring and open wiring on insulators. Note that these open wiring methods are without a protective outer jacket or enclosure and are extremely limited in application.

FIGURE 10-7 An example of an other than circular raceway is metal wireways.

FIGURE 10-8 Busway installation.

WIRING MATERIALS

Wiring materials are used to facilitate the installation of wiring methods. For example, *enclosures/boxes* are necessary for the installation of all types of wiring methods and equipment. Examples include cabinets, cutout boxes, meter socket enclosures, outlet boxes, device boxes, pull boxes, junction boxes, conduit bodies, fittings, and handhole enclosures (Figures 10–9a and b).

FIGURE 10–9   Outlet, device, and junction boxes as well as conduit bodies are covered in Chapter 3.

(a)

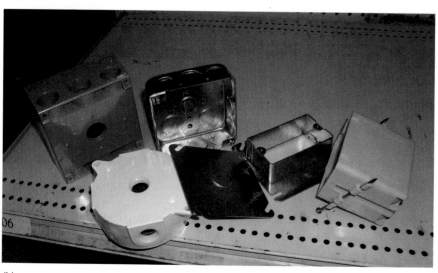

(b)

Support systems are a type of wiring material. Cable tray and messenger-supported wiring are a means of support for a given wiring method (Figure 10–10).

FIGURE 10-10   Cable tray is a support system for cables and/or raceways.

DidYouKnow?

Cable tray is a "wiring material" and not a wiring method; it is not a raceway. Cable tray is very simply a support method for raceways, cables, and conductors as permitted in Article 392.

GENERAL INFORMATION FOR ALL WIRING METHODS AND MATERIALS

Article 300 Wiring Methods

The first article in Chapter 3 is Article 300 Wiring Methods. General rules for *all* wiring methods and wiring materials are covered in Article 300. All electrical installations must conform to the general requirements in Article 300 unless modified by another article in Chapters 5, 6, or 7, Article 300 is the backbone of the "Build" chapter. General requirements for all wiring methods and materials are listed in this article rather than being repeated, which would be required for each raceway or cable assembly article.

Key provisions of Article 300 Wiring Methods include the following:

300.4 Protection Against Physical Damage This section requires that all conductors subject to physical damage be protected. Specific requirements are given for wiring methods through framing members and for ungrounded (hot) conductors 4 AWG (American Wire Gauge) or larger entering a cabinet, box, enclosure, or raceway.

300.5 Underground Installations Specific requirements for all underground installations of direct burial cables and raceways are illustrated in this section and the minimum cover requirement table.

300.11 Securing and Supporting This section requires that all raceways, cable assemblies, boxes, cabinets, and fittings be securely fastened in place.

Specific requirements are included for wiring located within floor, ceiling, and roof assemblies. Specific support requirements for raceways or cable assemblies for support spacing is included in the raceway or cable assembly article.

300.14 Length of Free Conductors at Outlets, Junctions, and Switch Points   This section requires a minimum length of conductor from boxes and enclosures to allow for splicing or device termination.

300.15 Boxes, Conduit Bodies, or Fittings—Where Required   This section requires that, in general, when raceways or cable assemblies are used, a box or conduit body (an LB for example) be installed at each conductor splice point, outlet point, switch point, junction point, or termination point.

300.21 Spread of Fire or Products of Combustion   This section requires that when wiring methods or materials penetrate an opening through fire resistant walls, partitions, floors, or ceilings, that the opening be fire-stopped through the use of approved methods to maintain the fire resistance rating.

300.22 Wiring in Ducts, Plenums, and Other Air-Handling Spaces   This section specifically addresses permitted wiring methods and materials in ducts, plenums, and other air-handling spaces.

Code users must become extremely familiar with the general requirements listed in Article 300 as they apply to every electrical installation regardless of the materials used.

Article 300 is separated into two logical parts as follows:

Part I General Requirements
- Voltage and Temperature Limitations
- Single Conductors
- Grouping of Conductors of the Same Circuit
- Paralleled Conductors of Different Systems
- Protection of Conductors
- Underground Installations
- Protection from Corrosion/Deterioration
- Raceways Exposed to Different Temperatures
- Electrical Continuity of Raceways
- Securing and Supporting
- Mechanical and Electrical Continuity of Conductors
- Where Boxes, Conduit Bodies, and Fittings Are Required
- Transition from Raceway or Cable to Open or Concealed Wiring
- Number and Size of Conductors in a Raceway
- Raceway Installations, General
- Supporting Conductors in Vertical Raceways
- Induced Currents in Metal Enclosures or Raceways
- Spread of Fire or Products of Combustion
- Wiring in Ducts, Plenums, and Other Air-Handling Spaces
- Panels Designed to Allow Access

DidYouKnow?

The answer to a general question about wiring methods or wiring materials will be found in Article 300. The answer to a question which deals with the application of a specfic wiring method such as "uses not permitted for Type MC Cable" will be found in the individual Article.

- Required Covers
- Conductors of Different Systems
- Conductor Bending Radius
- Protection Against Induction Heating
- Aboveground Wiring Methods
- Braid-Covered Insulated Conductors
- Insulation Shielding
- Underground Installations
- Moisture/Mechanical Protection for Metal-Sheathed Cables

CONDUCTORS

Article 310 Conductors for General Wiring

All general requirements for conductors in cable assemblies, raceways, open wiring, or support systems are contained in Article 310 Conductors for General Wiring. Flexible cords and cables are covered in Article 400 and fixture wires are covered in Article 402.

The general requirements of Article 310 Conductors for General Wiring cover the following:

Type designations, such as THWN or XHHW, are part of the required marking for all conductors. The letters and suffixes used in the type designation explain the physical properties of the conductor (Figure 10–11). These type designations are extremely important to the user of the code. Table 310.13

DidYouKnow?

Article 310 covers general requirements for conductors and their type designations, insulations, markings, mechanical strengths, ampacity ratings, and uses.

FIGURE 10–11 Article 310 covers general requirements for conductors and their type designations, insulations, markings mechanical, strengths, ampacity ratings, and uses.

provides the trade name, maximum operating temperatures, applications, insulation, thickness of insulation, and outer covering for the individual type conductors. Examples of letters used on conductors as type designations and their meaning are as follows:

T   Thermoplastic insulation

R   Thermoset insulation

S   Silicone (Thermoset) insulation

X   Cross-linked synthetic polymer insulation

Z   Modified tetrafluoroethylene insulation

U   Underground use

L   Lead sheath

N   Nylon jacket

W   Moisture resistant

H   75°C rated (Note: the lack of "H" usually indicates 60°C rating)

HH   90°C rated

-2   The suffix "-2" designates continuous 90°C rating, wet or dry

Insulation requirements for all conductors specify the type and thickness of all conductor insulation. Types of conductor insulation include but are not limited to TW, THW, THHW, THHN, RHH, RHW, and XHHW. The marking of conductors requires that all conductors and cables be marked to indicate maximum-rated voltage, proper type designation, manufacturer's name/trademark, and the size of the conductor in accordance with the circular mil area (cmil) or the American Wire Gauge (AWG).

Mechanical strengths of conductors are addressed by requirements for minimum size, thickness of insulation, outer coverings, and permitted applications.

Ampacity ratings for all types and uses of conductors are included in the ampacity tables in Article 310. Each of these ampacity tables address different installation possibilities for all types of conductors in all ranges of temperature limitations and installation methods. Table 310.16, however, is the most frequently used ampacity table. Section 310.15 provides specific requirements for conductor ampacity and the corrections for exceeding the number of current-carrying conductors permitted in the tables. The type designation letters, such as type THWN, are used to determine the ampacity of the conductor using the proper table.

Uses for all type designations of conductors are detailed in Table 310.13. For example, type THHW is permitted for use in dry locations with a temperature limitation of 90°C and in wet locations with a temperature limitation of 75°C.

Key provisions of Article 310 Conductors for General Wiring include the following:

310.3 Stranded Conductors   Requires all conductors, 8 AWG and larger, installed in raceways to be stranded.

310.4 Conductors in Parallel   This section requires that, in general, only conductors 1/0 AWG or larger shall be permitted to be installed in parallel.

310.8 Locations   This section permits all conductors to be installed in dry locations and provides a list of type designations permitted in damp and wet locations.

DidYouKnow?

All electrical installations will include conductors that will be installed in cable assemblies, raceways, or support methods. All conductors are subject to the general requirements of Article 310.

310.10 Temperature Limitation of Conductors This section requires that the temperature limitation determined by the type designation for all conductors shall not be exceeded.

310.11 Marking This section details required marking information and the methods permitted for marking.

310.15 Ampacities for Conductors Rated 0–2000 Volts This section provides general information on conductor ampacity and specific requirements for ampacity adjustment.

Code users must become extremely familiar with the general requirements listed in Article 310 as they apply to every electrical installation regardless of the conductors used. Article 310 is not separated into parts.

Article 310 Conductors for General Wiring

- Conductors
- Stranded Conductors
- Conductors in Parallel
- Minimum Size of Conductors
- Locations
- Corrosive Conditions
- Marking
- Conductor Identification
- Temperature Limitation of Conductors
- Aluminum Conductors
- Conductor Construction and Application
- Ampacities

CABINETS, BOXES, FITTINGS, AND METER SOCKET/HANDHOLE ENCLOSURES

In order to facilitate the installation of all types of wiring methods, the need arises for wiring materials. Cabinets, boxes, fittings, and meter socket/ handhole enclosures are essential wiring materials to all electrical installations (Figures 10–12a–c).

There are two articles in Chapter 3 that cover these wiring materials:

Article 312 Cabinets, Cutout Boxes, and Meter Socket Enclosures

Article 312 provides general rules for the installation and construction of cabinets, cutout boxes, and meter socket enclosures.
Key provisions of this article include the following:

312.2 Damp and Wet Locations This section requires that all enclosures are designed and installed to prevent moisture or water from entering and/or accumulating within the cabinet, cutout box, or meter socket enclosure.

DidYouKnow?

In accordance with the definition of "panelboard" in Article 100, a panelboard is designed to be installed in a cabinet or cutout box; specific information and requirements for cabinets and cutout boxes are located in Article 312.

FIGURE 10–12 Cabinets (containing panelboards), meter sockets, and handhole enclosures are wiring materials located in Chapter 3.

(a)

(b)

(c)

312.6 Deflection of Conductors This section requires that sufficient space be provided for bending and installation of conductors.

312.8 Enclosures for Switches or Overcurrent Devices This section prohibits enclosures for overcurrent devices from being used for other purposes unless adequate space has been provided.

Article 312 is separated into two parts as follows:

Part I Installation

- Damp, Wet, or Hazardous Locations
- Position in Wall
- Repairing Plaster and Drywall or Plasterboard
- Openings to be Closed
- Deflection of Conductors
- Space in Enclosures
- Enclosures for Switches or Overcurrent Devices
- Side or Back Wiring Spaces or Gutters

Part II Construction Specifications

- Material
- Spacing

Article 314 Outlet, Device, Pull and Junction Boxes; Conduit Bodies; Fittings; and Handhole Enclosures

Article 314 provides general rules for the installation and construction of outlet, device, pull and junction boxes; conduit bodies; fittings; and handhole enclosures (Figures 10–13a and b).

Key provisions of this article include the following:

314.16 Number of Conductors in Outlet, Device, and Junction Boxes, and Conduit Bodies This section, along with the table provided, requires that all boxes and conduit bodies are of sufficient size to provide free space for enclosed conductors. Note that this section covers installations where all conductors enclosed are 6 AWG or smaller.

314.23 Supports This section provides minimum requirements for the support of all boxes and enclosures.

314.28 Pull and Junction Boxes and Conduit Bodies This section provides minimum requirements for the size of all pull/junction boxes and conduit bodies. Note that this section covers installations where conductors 4 AWG or larger are enclosed.

314.29 Boxes, Conduit Bodies, and Handhole Enclosures to Be Accessible This section prohibits boxes and conduit bodies from being concealed. Note that a box installed above a lay-in type drop ceiling would be considered accessible.

314.71 Size of Pull and Junction Boxes This section exists in Part IV of Article 314; it is titled "Pull and Junction Boxes for Use on Systems Over 600 Volts, Nominal." This requires that all pull/junction boxes containing

DidYouKnow?

Article 314 contains information and requirements for the installation and use of all boxes, conduit bodies, and handhole enclosures.

FIGURE 10–13 Pull and device boxes.

(a)

(b)

conductors at over 600-volts be sized larger to accommodate the bending radius of high-voltage cables.

Article 314 is subdivided into four logical parts as follows:

Part I Scope and General

- Round Boxes
- Nonmetallic Boxes
- Metal Boxes

Part II Installation

- Damp/Wet/Hazardous Locations
- Number of Conductors in Outlet/Device/Junction Boxes and Conduit Bodies
- Conductors Entering Boxes/Fittings and Conduit Bodies
- Boxes Enclosing Flush Devices
- Boxes in Walls/Ceilings
- Repairing Plaster/Drywall/Plasterboard
- Exposed Surface Extensions
- Supports
- Depth of Outlet Boxes
- Covers and Canopies
- Outlet Boxes
- Pull/Junction and Conduit Bodies (Size)
- Accessibility of Boxes/Conduit Bodies/Handhole Enclosures
- Handhole Enclosures

Part III Construction Specifications

- Metal Boxes/Fittings and Conduit Bodies Covers
- Bushings
- Nonmetallic Boxes
- Marking

Part IV Pull and Junction Boxes for Use on Systems Over 600 Volts, Nominal

- General
- Pull/Junction Boxes (Size)
- Construction and Installation Requirements

CABLE ASSEMBLIES AND RACEWAYS

Similar Article Layout for Usability

All *cable assembly* and *circular raceway* articles share a common article layout and section numbering system to provide the Code user with a consistent, easy-to-use format. Many of the "other than circular" raceways have also adopted this common format. Once familiar with the common format of these articles, the Code user can quickly and accurately move through the many different wiring methods in Chapter 3 to find information and requirements. For example, the .10 section of each article is "Uses Permitted." A Code user fa-

DidYouKnow?

To facilitate the installation of large conductors in a raceway, a straight through "pull box" is typically installed. 314.28 (A)(1) requires that the length of the pull box be not less than eight times the trade size of the largest raceway.

DidYouKnow?

The common numbering format of Chapter 3 raceway and cable assembly Articles is designed to allow the Code user to quickly and accurately determine permitted wiring methods for a given installation.

miliar with this common numbering scheme could quickly and accurately determine the wiring methods permitted for a particular installation.

Within the common format, there is specific section numbering that may not apply to all wiring methods. For example, the .28 section is reserved for "Reaming and Threading." This section will exist only in metal raceway articles where reaming and threading are necessary for installation. The common article layout and section numbering is as follows:

Common Article Format
Article 3XX

Part I General

3XX.1	Scope
3XX.2	Definition/s
3XX.3	Other Articles
3XX.6	Listing Requirements

Part II Installation

3XX.10	Uses Permitted
3XX.12	Uses Not Permitted
3XX.15	Exposed Work
3XX.17	Through or Parallel to Framing Members
3XX.19	Clearances
3XX.20	Size
3XX.23	In Accessible Attics
3XX.24	Bends/Made
3XX.26	Bends/Number in One Run
3XX.28	Reaming and Threading
3XX.30	Securing and Supporting
3XX.40	Boxes/Fittings
3XX.42	Couplings and Connectors
3XX.46	Bushings
3XX.48	Joints
3XX.56	Splices and Taps
3XX.60	Grounding/Bonding
3XX.80	Ampacity

Part III Construction Specifications

3XX.100	Construction
3XX.104	Conductors
3XX.108	Equipment Grounding
3XX.112	Insulation
3XX.116	Sheath/Jacket
3XX.120	Marking
3XX.130	Standard Lengths

DidYouKnow?

Cable assemblies and raceways are separated and grouped together in Chapter 3 to logically organize, present, and list them according to their physical characteristics.

Cable Assemblies

There are 11 types of *cable assemblies* recognized as acceptable wiring methods in Chapter 3. All cable assembly articles are listed in alphabetical order as follows:

Article 320 Armored Cable: Type AC

Article 322 Flat Cable Assemblies: Type FC

Article 324 Flat Conductor Cable: Type FCC

Article 326 Integrated Gas Spacer Cable: Type IGS

Article 328 Medium Voltage Cable: Type MV

Article 330 Metal-Clad Cable: Type MC

Article 332 Mineral Insulated, Metal-Sheathed Cable: Type MI

Article 334 Nonmetallic-Sheathed Cable: Types NM, NMC, and NMS

Article 336 Power and Control Tray Cable: Type TC

Article 338 Service Entrance Cable: Types SE and USE

Article 340 Underground Feeder and Branch Circuit Cable: Type UF

Raceways, Circular Metal Conduit

There are four types of *circular metal conduits* recognized as an acceptable wiring method in Chapter 3. They are separated into *rigid type* conduits and *flexible type* conduits as follows:

Article 342 Intermediate Metal Conduit: Type IMC

Article 344 Rigid Metal Conduit: Type RMC

Article 348 Flexible Metal Conduit: Type FMC (Figure 10–14)

Article 350 Liquidtight Flexible Metal Conduit: Type LFMC

FIGURE 10–14 Flexible Metal Conduit, Type FMC.

Raceways, Circular Nonmetallic Conduit

There are four types of *circular nonmetallic conduits* recognized as an acceptable wiring method in Chapter 3. They are separated into *rigid type* conduits and *flexible type* conduits as follows:

Article 352 Rigid Nonmetallic Conduit: Type RNC

Article 353 High-Density Polyethylene Conduit: Type HDPE Conduit

Article 354 Nonmetallic Underground Conduit with Conductors: Type NUCC

Article 355 Reinforced Thermosetting Resin Conduit: Type RTRC

Article 356 Liquidtight Flexible Nonmetallic Conduit: Type LFNC

Raceways, Circular Metallic Tubing

There are two types of *circular metallic tubing* recognized as an acceptable wiring method in Chapter 3. Note that Electrical Metallic Tubing, type EMT, is commonly called "Thinwall Conduit" but is designated as "Tubing" in the *NEC*®. There is a single *rigid type* tubing and a single *flexible type* tubing as follows:

Article 358 Electrical Metallic Tubing: Type EMT

Article 360 Flexible Metallic Tubing: Type FMT

Raceways, Circular Nonmetallic Tubing

There is a single type of *circular nonmetallic tubing* recognized as an acceptable wiring method, and is as follows:

Article 362 Electrical Nonmetallic Tubing: Type ENT

Factory-Assembled Power Distribution Systems

The *NEC*® recognizes two types of *factory-assembled power distribution systems* as recognized wiring methods in Chapter 3. These wiring methods, *busway* and *cablebus*, are preassembled and bolted together for a complete installation in the field. These systems can allow for a disconnecting means and overcurrent protection to be installed anywhere along the installation of busway and cablebus, providing an easy means for power distribution. The two articles for busway and cablebus numerically separate the "Circular Raceways" from the "Other than Circular Raceways," and are as follows:

Article 368 Busways

Article 370 Cablebus

Raceways Other Than Circular

There are ten types of *other than circular raceways* recognized as an acceptable wiring method in Chapter 3. These are as follows:

Article 366 Auxiliary Gutters

Article 372 Cellular Concrete Floor Raceways

Article 374 Cellular Metal Floor Raceways

Article 376 Metal Wireways

Article 378 Nonmetallic Wireways

Article 380 Multioutlet Assemblies

Article 384 Strut-Type Channel Raceway

Article 386 Surface Metal Raceways

Article 388 Surface Nonmetallic Raceways

Article 390 Underfloor Raceways

DidYouKnow?

Article 392 Cable Trays and Article 396 Messenger-Supported Wiring are not raceways or wiring methods. They are support systems.

Surface-Mounted Nonmetallic Branch Circuit Extension

This type of wiring method is primarily limited to use only from an existing outlet in a residential or commercial occupancy not more than three floors above grade. This wiring method was used primarily in older electrical installations to allow the surface mounting of additional receptacle outlets. The 2008 *NEC®* includes provisions to allow a "concealable nonmetallic extension" to be installed on walls or ceilings covered with paneling, tile, joint compound, or similar material. This type of wiring method is included in the following article:

Article 382 Nonmetallic Extensions

Support Systems for Cables/Raceways

There are two types of systems recognized as acceptable for the support of wiring methods in Chapter 3. *Cable tray* is permitted under specified conditions to support single conductors, cable assemblies, and raceways. *Messenger-supported wiring* is as the name implies, an exposed wiring support system using a messenger wire to support insulated conductors. This system is permitted to support only cable assemblies or conductors listed in Table 396.10(A). The two types of systems are as follows:

Article 392 Cable Trays

Article 396 Messenger-Supported Wiring

Open-Type Wiring Methods

There are two types of open-type wiring methods permitted in Chapter 3. These open wiring methods are extremely limited in use, and are as follows

Article 394 Concealed Knob and Tube Wiring

Article 398 Open Wiring on Insulators

KEY WORDS/CLUES FOR CHAPTER 3, "BUILD"

Wiring Methods

- General questions for wiring method installation
- Conductors, types, uses, ampacity
- Cable assemblies, all types
- Conduits, all types
- Tubing, all types
- Other raceways, all types
- Installation of all wiring methods
- Support of all wiring methods

- Construction of all wiring methods
- Uses permitted or not permitted, all wiring methods

Wiring Materials

- General questions for wiring materials
- Cabinets
- Cutout boxes
- Meter socket enclosures
- Outlet, device, pull and junction boxes
- Conduit bodies
- Handhole enclosures
- Support systems, cable tray and messenger-supported wiring
- Construction of wiring materials
- Installation of wiring materials
- Support of wiring materials

THINK "BUILD" AND GO TO CHAPTER 3

Refer to Chapter 3 when your question or need within the *NEC*® deals with:
- Wiring methods, cable assemblies, raceways
- Conductors, type designation, use, ampacity
- Installation of wiring methods
- Wiring materials, enclosures of all types
- Installation, use, and size of wiring materials
- Any question on the physical installation of wiring methods and materials
- Support systems for raceways, conductors, or cable assemblies

SUMMARY

Chapter 3, in accordance with Section 90.3, applies generally to all electrical installations. Using the Codeology method, this chapter is given the nickname of the Build chapter, due to the scope covered. Chapter 3 is a "hands-on" chapter. All of the material covered is to be physically installed; it is the means by which electrical current will get from the source of power to the last outlet in the electrical distribution system.

The *NEC®* title for Chapter 3 is "Wiring Methods and Materials." From this title, the scope of this chapter can be described as "Information and Rules on *Wiring Methods and Materials* for Electrical Installations." Chapter 3 covers the entire electrical distribution system from the service point (connection to the utility) to the last recep-

tacle or other outlet in the electrical system. All of the wiring methods and materials used to distribute electrical energy from the source to the last outlet are covered in Chapter 3. Chapter 3, in accordance with its scope of "Wiring Methods," provides detailed requirements for all conductors, raceways, cable assemblies, and other recognized wiring methods. Chapter 3, in accordance with its scope of "Wiring Materials," provides detailed requirements for all enclosures, boxes, conduit bodies, and support systems.

See Figure 10–15 on page 178 for an example of how Chapter 3 along with Chapters 1, 2, and 4 build the foundation or backbone of all electrical installations.

REVIEW QUESTIONS

1. Does Chapter 3 of the *NEC®* apply to all electrical installations covered by the *NEC®*?

2. Chapter 3 is subdivided into how many articles?

3. The scope of Chapter 3 of the *NEC®* is separated into two areas covering wiring _____ and wiring _____.

4. What part of which article in Chapter 3 would address the installation of a pull box containing conductors rated at 13,200 volts?

5. Chapter 3 of the *NEC®* addresses wiring methods, which would include circular raceways. What part of which article would apply if one were installing type RMC conduit?

6. Does 300.22 provide any information to aid the Code user to understand what "other space" is in relation to a duct or plenum?

7. Which two articles in Chapter 3 cover requirements for support systems for cables and/or raceways?

8. How many parts is Article 340 subdivided into?

9. Does Part III of Article 340 apply to the installation of type UF cable?

10. Which two articles in Chapter 3 cover requirements for factory-assembled power distribution systems?

FIGURE 10–15 Chapter 3 will apply generally in all electrical installations.

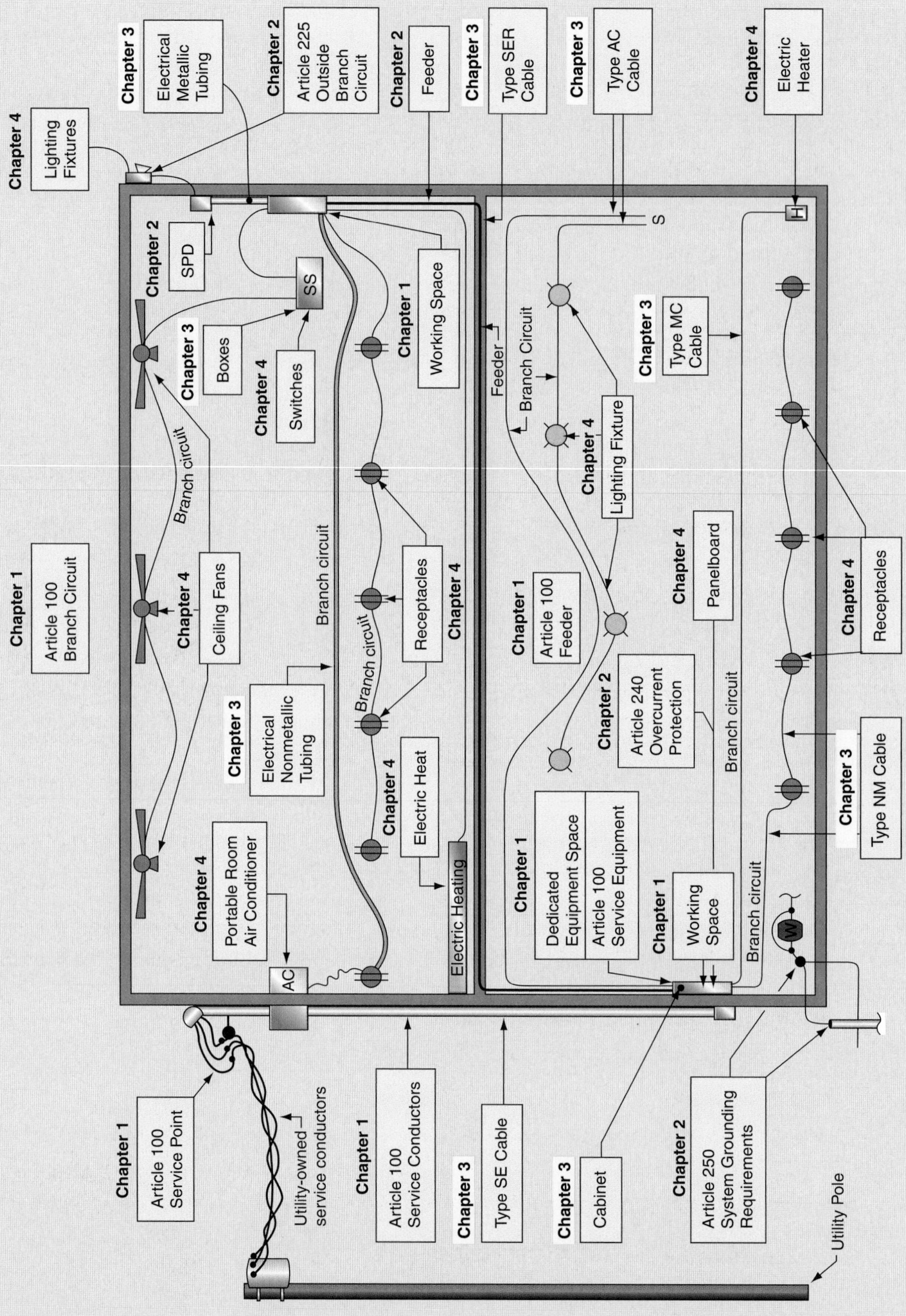

Chapter 4
Lighting Fixtures

Chapter 3
Electrical Metallic Tubing

Chapter 2
Article 225 Outside Branch Circuit

Chapter 2
Feeder

Chapter 3
Type SER Cable

Chapter 3
Type AC Cable

Chapter 4
Electric Heater

Chapter 1
Article 100 Branch Circuit

Chapter 2
SPD

Chapter 3
Boxes

Chapter 4
Ceiling Fans

Chapter 4
Switches

SS

Chapter 1
Working Space

Branch circuit

Branch circuit

Branch circuit

Branch circuit

Feeder

Branch Circuit

Chapter 4
Lighting Fixture

Chapter 3
Type MC Cable

Chapter 3
Electrical Nonmetallic Tubing

Chapter 4
Receptacles

Chapter 4
Electric Heat

Chapter 4
Portable Room Air Conditioner

AC

Electric Heating

Chapter 1
Dedicated Equipment Space
Article 100 Service Equipment

Chapter 1
Working Space

Chapter 1
Article 100 Feeder

Chapter 2
Article 240 Overcurrent Protection

Chapter 4
Panelboard

Branch circuit

Chapter 4
Receptacles

Chapter 3
Type NM Cable

S

H

W

Chapter 1
Article 100 Service Point

Utility-owned service conductors

Chapter 1
Article 100 Service Conductors

Chapter 3
Type SE Cable

Chapter 3
Cabinet

Chapter 2
Article 250 System Grounding Requirements

Branch circuit

Utility Pole

PRACTICE PROBLEMS

Using the Codeology Method

The following questions and steps to find the answer are designed to illustrate the Codeology method. While the steps to find your answer may seem lengthy, they are designed to illustrate the thought process to find the answer. The easiest way to make the Codeology method work is to silently "talk to yourself." Walk through these steps by silently talking to yourself and Codeology will be a natural response for quickly and accurately finding needed information in the *NEC*®.

Read the following questions and follow step by step using your codebook and the Codeology Method.

1. When permanent barriers are installed in a pull box or junction box, is each separate compartment considered a separate box?

Step #1

Qualify the question or need, look for key words
    This question is about a pull box or junction box.
    Key words: "*pull box or junction box*"
    Think wiring materials, think BUILD

Step #2

Go to the table of contents, Chapter 3, the Build chapter
    Find the correct article   **314 Outlet, Device, Pull, and Junction Boxes; Conduit Bodies; Fittings; and Handhole Enclosures**

Step #3

Further qualify your question or need and get in the right part of the article
    Key words: "*permanent barriers are installed*"
    Look in **Part II Installation**

Step #4

Read each section title until the correct section is found
    Identify correct section or subdivision and find answer
    Correct section/subdivision

Section   **314.28 Pull and Junction Boxes and Conduit Bodies**
First-Level Subdivision   **(D) Permanent Barriers**

2. Is flat conductor cable permitted for use as a feeder?

Step #1

Qualify the question or need, look for key words
    This question is about the permitted use of a cable assembly.
    Key words: "*flat conductor cable*"
    Think wiring methods, think BUILD

Step #2

Go to the table of contents, Chapter 3, the Build chapter
    Find the correct article   **324 Flat Conductor Cable: Type FCC**

Step #3

Further qualify your question or need and get in the right part of the article
    Key words: "*permitted use*"
    Look in   **Part II Installation**

Step #4

Read each section title until the correct section is found
    Identify correct section or subdivision and find answer
    Correct section/subdivision
Section   **324.10 Uses Permitted**
First-Level Subdivision   **(A) Branch Circuits**

3. What is the maximum number of current-carrying conductors at any cross-section permitted in a metal wireway before ampacity derating factors must be applied?

Step #1

Qualify the question or need, look for key words
    This question is about a metal wireway.
    Key words: "*permitted use of a metal wireway*"
    Think wiring methods, think BUILD

Step #2

Go to the table of contents, Chapter 3, the Build chapter

Find the correct article   **376 Metal Wireways**

Step #3

Further qualify your question or need and get in the right part of the article

Key words: *"permitted use"*

Look in   **Part II Installation**

Step #4

Read each section title until the correct section is found

Identify correct section or subdivision and find answer

Correct section/subdivision

Section   **376.22 Number of Conductors (B) Adjustment Factors**

---

4. Type TBS conductors are only permitted to be applied in _____.

Step #1

Qualify the question or need, look for key words

This question is about the permitted use of a conductor.

Key words: *"type TBS conductors"*

Think wiring methods, think BUILD

Step #2

Go to the table of contents, Chapter 3, the Build chapter

Find the correct article   **310 Conductors for General Wiring**

Step #3

Further qualify your question or need and get in the right part of the article

Key words: *"permitted application"*

Article 310 is not separated into parts.

Step #4

Read each section title until the correct section is found

Identify correct section or subdivision and find answer

Correct section/subdivision

Section   **310.13 Conductor Constructions and Applications**

Table   **Table 310.13 Conductor Applications and Insulations**

---

5. Cables and insulated conductors installed in underground raceways shall be listed for use in _____ locations.

Step #1

Qualify the question or need, look for key words

This question is about an underground raceway installation.

Key words: *"insulated conductors installed in underground raceways shall be listed for"*

Think wiring methods, think BUILD

This is a general question about an underground raceway installation.

Step #2

Go to the table of contents, Chapter 3, the Build chapter

Find the correct article   **300 Wiring Methods**

Step #3

Further qualify your question or need and get in the right part of the article

Key words: *"installed in underground raceways"*

Look in   **Part I General Requirements**

Step #4

Read each section title until the correct section is found

Identify correct section or subdivision and find answer

Correct section/subdivision

Section   **300.5 Underground Installations**

First-Level Subdivision   **(B) Wet Locations**

---

6. Are cable trays permitted to be installed in hoistways?

Step #1

Qualify the question or need, look for key words

This question is about the use of cable trays.

Key words: *"cable trays permitted to be installed in hoistways"*

Think wiring materials, think BUILD

Step #2

Go to the table of contents, Chapter 3, the Build chapter

Find the correct article   **392 Cable Trays**

Step #3

Further qualify your question or need and get in the right part of the article

Key words: *"cable trays permitted to be installed in hoistways"*

Article 392 is not separated into parts.

Step #4

Read each section title until the correct section is found

Identify correct section or subdivision and find answer

Correct section/subdivision

Section   **392.4 Uses Not Permitted**

---

7. In general when paralleling conductors the smallest size permitted is _____ AWG.

Step #1

Qualify the question or need, look for key words

This question is about paralleling conductors.

Key words: *"when paralleling conductors"*

Think wiring methods, think BUILD

Step #2

Go to the table of contents, Chapter 3, the Build chapter

Find the correct article   **310 Conductors for General Wiring**

Step #3

Further qualify your question or need and get in the right part of the article

Key words: *"when paralleling conductors"*

Article 310 is not separated into parts.

Step #4

Read each section title until the correct section is found

Identify correct section or subdivision and find answer

Correct section/subdivision

Section   **310.4 Conductors in Parallel**

---

8. What is the standard length, in feet, of intermediate metal conduit?

Step #1

Qualify the question or need, look for key words

This question is about Intermediate Metal Conduit.

Key words: *"intermediate metal conduit"*

Think wiring methods, think BUILD

Step #2

Go to the table of contents, Chapter 3, the Build chapter

Find the correct article   **342 Intermediate Metal Conduit: Type IMC**

Step #3

Further qualify your question or need and get in the right part of the article

Key words: *"standard length"*

Look in   **Part III Construction Specifications**

Step #4

Read each section title until the correct section is found

Identify correct section or subdivision and find answer

Correct section/subdivision

Section   **342.130 Standard Lengths**

---

9. Type USE cable is specifically identified for _____ use.

Step #1

Qualify the question or need, look for key words

This question is about USE cable.

Key words: *"type USE cable"*

Think wiring methods, think BUILD

Step #2

Go to the table of contents, Chapter 3, the Build chapter

Find the correct article   **338 Service-Entrance Cable: Types SE and USE**

Step #3

Further qualify your question or need and get in the right part of the article

Key words: *"identified use of type USE cable"*

Look in   **Part I General**

Step #4

Read each section title until the correct section is found

Identify correct section or subdivision and find answer

Correct section/subdivision

Section   **338.2 Definitions**

---

10. In general, are circuits of different voltages/systems permitted to be installed in the same raceway or cable?

Step #1

Qualify the question or need, look for key words

This is a general question for wiring methods.

Key words: *"circuits of different voltages/systems permitted to be installed in the same raceway or cable"*

Think wiring methods, think BUILD

This is a general question about wiring methods.

Step #2

Go to the table of contents, Chapter 3, the Build chapter

Find the correct article   **300 Wiring Methods**

Step #3

Further qualify your question or need and get in the right part of the article

Key words: *"circuits of different voltages/systems"*

Look in   **Part I General Requirements**

Step #4

Read each section title until the correct section is found

Identify correct section or subdivision and find answer

Correct section/subdivision

Section   **300.3 Conductors**

First-Level Subdivision   **(C) Conductors of Different Systems**

Second-Level Subdivision   **(1) 600 Volts, Nominal, or Less**

---

11. In general electrical metallic tubing must be securely fastened in place every _____ feet and within _____ feet of each outlet box, junction box, device box, cabinet, conduit body, or other tubing termination.

Step #1

Qualify the question or need, look for key words

This is a question about supporting EMT.

Key words: *"securely fastened in place"*

Think wiring methods, think BUILD

This is a question about the installation of EMT.

Step #2

Go to the table of contents, Chapter 3, the Build chapter

Find the correct article   **358 Electrical Metallic Tubing: Type EMT**

Step #3

Further qualify your question or need and get in the right part of the article

Key words: *"securely fastened in place"*

Look in   **Part II Installation**

Step #4

Read each section title until the correct section is found

Identify correct section or subdivision and find answer

Correct section/subdivision

Section   **358.30 Securing and Supporting**

First-Level Subdivision   **(A) Securely Fastened**

---

12. The bending radius of the curve of the inner edge of any bend in the interlocked type of Metal-Clad cable shall not be less than _____ times the external diameter of the metallic sheath.

Step #1

Qualify the question or need, look for key words

This is a question about metal-clad cable.

Key words: *"bending, metal-clad cable"*

Think wiring methods, think BUILD

This is a question about the installation of metal-clad cable.

Step #2

Go to the table of contents, Chapter 3, the Build chapter

Find the correct article   **330 Metal-Clad Cable: Type MC**

Step #3

Further qualify your question or need and get in the right part of the article

Key words: *"bending radius, interlocked type MC cable"*

Look in   **Part II Installation**

Step #4

Read each section title until the correct section is found

Identify correct section or subdivision and find answer

Correct section/subdivision

Section   **330.24 Bending Radius**

First-Level Subdivision   **(B) Interlocked-Type Armor or Corrugated Sheath**

---

13. Busways shall be securely supported at intervals not exceeding _____ feet unless otherwise designed and marked.

Step #1

Qualify the question or need, look for key words

This question is about supporting busways.

Key words: *"securely support busways"*

Think wiring methods, think BUILD

Step #2

Go to the table of contents, Chapter 3, the Build chapter

Find the correct article   **368 Busways**

Step #3

Further qualify your question or need and get in the right part of the article

Key words: *"support busways"*

Look in   **Part II Installation**

Step #4

Read each section title until the correct section is found

Identify correct section or subdivision and find answer

Correct section/subdivision

Section   **368.30 Support**

---

14. When necessary to compensate for thermal expansion and contraction raceways shall be provided with _____ _____.

Step #1

Qualify the question or need, look for key words

This is a general question about raceways.

Key words: *"raceways, compensate for thermal expansion"*

Think wiring methods, think BUILD

Step #2

Go to the table of contents, Chapter 3, the Build chapter

Find the correct article   **300 Wiring Methods**

Step #3

Further qualify your question or need and get in the right part of the article

Key words: *"general question about raceways, not a specific type"*

Look in   **Part I General Requirements**

Step #4

Read each section title until the correct section is found

Identify correct section or subdivision and find answer

Correct section/subdivision

Section   **300.7 Raceways Exposed to Different Temperatures**

First-Level Subdivision   **(B) Expansion Fittings**

---

15. The maximum size copper-clad aluminum conductors in type UF cable is _____ AWG.

Step #1

Qualify the question or need, look for key words

This question is about type UF cable.

Key words: *"maximum size conductors"*

Think wiring methods, think BUILD

Step #2

Go to the table of contents, Chapter 3, the Build chapter

Find the correct article   **340 Underground Feeder and Branch-Circuit Cable: Type UF**

Step #3

Further qualify your question or need and get in the right part of the article

Key words: *"maximum size copper-clad aluminum conductors in type UF cable"*

Look in   **Part III Construction Specifications**

Step #4

Read each section title until the correct section is found

Identify correct section or subdivision and find answer

Correct section/subdivision

Section   **340.104 Conductors**

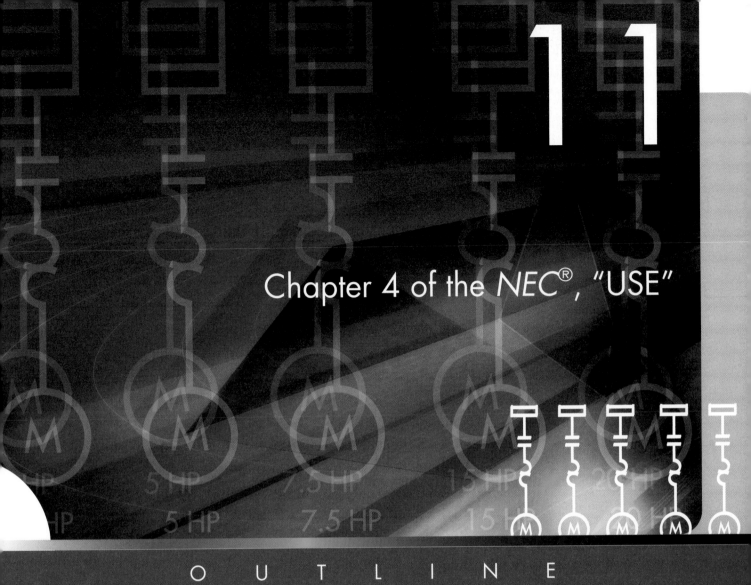

Chapter 4 of the *NEC*®, "USE"

OUTLINE

OBJECTIVES

After completing this unit, you should be able to:

1. Associate the Codeology title for *NEC®* Chapter 4 as "Use"
2. Identify the type of information and requirements dealing with the installation, control, and supply for utilization equipment contained in Chapter 4
3. Recognize key words and clues for locating answers in *NEC®* Chapter 4, the "Use" chapter
4. Identify exact sections, subdivisions, list items, etc. to justify answers for all questions referring to Chapter 4 articles
5. Recognize that Chapter 4 numbering is the 400-series
6. Recognize, recall, and become familiar with articles contained in Chapter 4

OVERVIEW

Chapter 4 of the *NEC®*, the "Use" chapter, provides rules and information on electrical equipment for general use. Special equipment is addressed in Chapter 6, in accordance with 90.3.

All equipment located in Chapter 4 is dedicated to the use of electrical energy. In Chapter 2 of the *NEC®*, a general electrical installation is *planned*. In Chapter 3 of the *NEC®*, a general electrical installation is *built,* getting electrical energy from the source to the load/s. In Chapter 4 of the *NEC®*, the use or consumption of electrical energy is addressed, including utilization equipment and all other equipment necessary. Electrical equipment that *uses* electrical energy performs a task or provides a service for the consumer. For example, lighting fixtures illuminate our homes, electrical space heaters provide heat, and an air conditioner keeps the home cool.

NEC® CHAPTER 4 "USE"

While Chapter 4 is identified as the USE chapter, not all of the equipment in this chapter *uses* electrical energy. However, all of the equipment in Chapter 4 plays a major role in the use of electrical energy. The following examples are listed to help explain the makeup of Chapter 4, the USE chapter:

- Lighting fixtures, appliances, heating equipment, motors, and air conditioning/refrigeration equipment all use electrical energy.
- Flexible cords and cables are necessary for the connection of utilization equipment, such as appliances to an electrical outlet.
- Fixture wires are necessary for the wiring of lighting fixtures.
- Panelboards, switchboards, industrial control panels and switches are necessary to provide control and overcurrent protection for all conductors supplying the end-use equipment.
- Receptacle outlets are necessary for the use of appliances and other loads.
- Switches are necessary to control lighting and all other loads.
- Generators are necessary to provide a power source for emergency, legally required standby, and optional standby systems. Other sources of power, such as solar and fuel cell systems are considered special and are in Chapter 6 of the *NEC*®.
- Transformers are necessary for the use of electrical energy. Transformers provide the flexibility of creating a new system to allow the consumption of electrical energy at utilization voltages. For example, a service may be 277/480 volts, 3-phase, 4-wire to supply air conditioning and refrigeration equipment. A transformer is installed that creates a new system at 120/208 volts to allow electrical energy consumption at 120 volts for general receptacle outlets.
- Phase converters, capacitors, resistors, and reactors are all necessary to allow economical use of electrical energy.

There are 21 articles in Chapter 4, the 400-series. These articles address equipment for general use to facilitate the utilization of electrical energy. The requirements and information in this chapter are logically separated into seven different categories as shown in Table 11–1.

TABLE 11-1   Categories of *NEC*® Chapter 4

NEC® Title:	Equipment for General Use
Codeology Title:	USE
Chapter Scope:	Information and Rules on *Equipment for General Use* in Electrical Installations

Connection/Wiring of Utilization Equipment

Control of Utilization Equipment

Utilization Equipment

Generation of Power for Utilization Equipment

Transformation of Systems for Utilization Equipment

Storage of Power for Utilization Equipment

High Voltage, Equipment Over 600 Volts, Nominal

CATEGORIZATION OF CHAPTER 4

When using the Codeology method, the title for Chapter 4 is USE. This title encompasses all equipment that *uses* electrical energy as well as all associated equipment necessary to safely accomplish utilization. The seven categories outlined in Table 11–1 organize the various types of equipment in Chapter 4.

Connection/Wiring of Utilization Equipment

Flexible cords and cables are appropriately located in Chapter 4, the USE chapter. Flexible cords and cables are not permitted as a substitute for fixed wiring (Figure 11.1). This means that while cords and cables are permitted to supply utilization equipment, they are not permitted as "wiring methods." Chapter 3 is dedicated to wiring methods and wiring materials.

Fixture wires are also located in Chapter 4. Fixture wires are not permitted to serve as a branch circuit. They are permitted only for installation in lighting fixtures or associated equipment.

Electrical systems are installed in all occupancies to allow for the use of electrical equipment. Utilization equipment must be connected to a branch circuit. This connection to a branch circuit can be hardwired or cord and plug connected. Article 400 provides information and requirements on the use of cords and cables.

Lighting fixtures are called by the proper international term in the *NEC*®, which is luminaires. Article 402 provides information and requirements for the use of fixture wire and limits that use to lighting fixtures and associated equipment. The two articles which address the connection/wiring requirements of Chapter 4 are as follows:

- 400 Flexible Cords and Cables
  - Part I   General
  - Part II   Construction Specifications
  - Part III Portable Cables Over 600 Volts, Nominal
- 402 Fixture Wires

Control of Utilization Equipment

Switches are appropriately located in Chapter 4, the USE chapter. Switches are essential to provide control and protection (fused switches) for conductors and utilization equipment. *Receptacles, cord connectors,* and *attachment plugs* are also located in Chapter 4, the USE chapter. The connection and disconnection of appliances and other cord- and plug-connected utilization equipment would not be possible without these devices. In addition, *switchboards, panelboards,* and *industrial control panels* are located in this chapter. This equipment is essential to allow for the control and protection of all conductors and utilization equipment (Figures 11–2a, b, and c).

The *NEC*® requires control and protection of all conductors and equipment in an electrical installation. In some cases utilization equipment is controlled and protected through the same devices used to protect the branch circuit conductors. Not all electrical "equipment for general use" is located at the end of the branch circuit with the utilization equipment. Article 408 Panelboards and Switchboards and Article 409 Industrial

FIGURE 11-1

Flexible cord and cable facilitate the installation and frequent interchange of utilization equipment.

DidYouKnow?

Chapter 4 of the *NEC*® consists of Articles that address utilization equipment; the connection and control of utilization equipment; and, where necessary, the generation, transformation, and storage of energy for utilization equipment.

FIGURE 11-2 Receptacles, switches, panelboards, and switchboards are equipment for general use providing control and protection of utilization equipment.

(a)

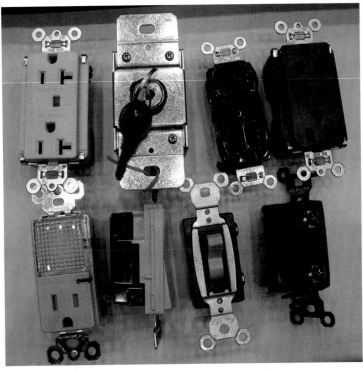

(b)

(c)

?

DidYouKnow?

Devices such as switches (Article 404) and receptacles (Article 406) are necessary for the control and the application of utilization equipment and are logically located in Chapter 4 of the *NEC*®.

Control Panels will provide control and protection of utilization equipment in most cases at the source of the feeder and branch circuits supplying the equipment. While Article 404 Switches may control only the utilization equipment or part of a branch circuit, Article 406 Receptacles, Cord Connectors and Attachment Plugs will be located at the utilization equipment providing control and flexibility (Figures 11–3a and b). The four articles which address the control requirements of Chapter 4 are as follows:

- 404 Switches
  - Part I   Installation
  - Part II Construction Specifications

FIGURE 11-3 Article 408 covers switchboards and panelboards.

(a)

(b)

DidYouKnow?

Panelboards and switch boards are required to legibly identify every circuit and circuit modification. This requirement is found in Article 408, Part I General in Section 408.4.

- 406 Receptacles, Cord Connectors and Attachment Plugs (Caps)
- 408 Switchboards and Panelboards
  - Part I   General
  - Part II   Switchboards (The Article 100 definition of switchboard is: *a large single panel, frame, or assembly of panels on which are mounted on the face, back, or both, switches, overcurrent and other protective devices, buses, and usually instruments. Switchboards are generally accessible from the rear as well as from the front and are not intended to be installed in cabinets.*)
  - Part III   Panelboards (The Article 100 definition of panelboard is: *a single panel or group of panel units designed for assembly in the form of a single panel, including buses and automatic overcurrent devices, and equipped with or without switches for control of light, heat, or power circuits; designed to be placed in a cabinet or cutout box placed in or against a wall, partition, or other support; and accessible only from the front.*)
  - Part IV   Construction Specifications
- 409 Industrial Control Panels
  - Part I   General
  - Part II   Installation
  - Part III   Construction Specifications

Utilization Equipment

Lighting fixtures (luminaires), appliances, space heating equipment, deicing/snow melting equipment, heat trace, motors, AC and refrigeration equipment are all appropriately located in Chapter 4, the USE chapter. This equipment utilizes or *uses* electrical energy (Figures 11–4a, b, c, d, and e).

The *NEC*® contains eight articles that specifically address the requirements of utilization equipment. These requirements for utilization equipment include: general requirements, installation requirements, control/protection requirements, disconnecting requirements, construction requirements, and many other equipment-specific requirements. *NEC*® requirements for special utilization equipment are located in Chapter 6 in conformance with Section 90.3. General utilization equipment in Chapter 4 of the *NEC*® falls into four basic categories as follows: lighting, appliances, heating, and motor equipment.

Lighting

- 410 Luminaires, Lampholders, and Lamps
  - Part I   General
  - Part II   Luminaire Locations
  - Part III   Provisions at Luminaire Outlet Boxes, Canopies, and Pans

FIGURE 11–4 Chapter 4 includes general utilization equipment including motors, air conditioners, appliances, lighting fixtures, and electric heat.

(a)

(b)

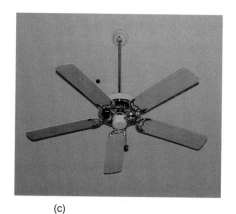

(c)

(d)

(e)

- Part IV   Luminaire Supports
- Part V   Grounding
- Part VI   Wiring of Luminaires
- Part VII   Construction of Luminaires
- Part VIII   Installation of Lampholders
- Part IX   Construction of Lampholders
- Part X   Lamps and Auxiliary Equipment
- Part XI   Special Provisions for Flush and Recessed Luminaires
- Part XII   Construction of Flush and Recessed Luminaires
- Part XIII   Special Provisions for Electric Discharge Lighting Systems of 1,000 Volts or Less
- Part XIV   Special Provisions for Electric Discharge Lighting Systems of More Than 1,000 Volts
- Part XV   Lighting Track
- Part XVI   Decorative Lighting and Similar Accessories
- 411 Lighting Systems Operating at 30 Volts or Less

Appliances

- 422 Appliances
  - Part I   General
  - Part II   Installation
  - Part III   Disconnecting Means
  - Part IV   Construction
  - Part V   Marking

Heating

- 424 Fixed Electric Space Heating Equipment
  - Part I   General
  - Part II   Installation
  - Part III   Control and Protection of Fixed Electric Space Heating Equipment
  - Part IV   Marking of Heating Equipment
  - Part V   Electric Space Heating Cables
  - Part VI   Duct Heaters
  - Part VII   Resistance-Type Boilers
  - Part VIII   Electrode-Type Boilers
  - Part IX   Electric Radiant Heating Panels and Heating Panel Sets
- 426 Fixed Outdoor Electric De-icing and Snow Melting Equipment
  - Part I   General
  - Part II   Installation
  - Part III   Resistance Heating Elements
  - Part IV   Impedance Heating
  - Part V   Skin Effect Heating
  - Part VI   Control and Protection
- 427 Fixed Electric Heating Equipment for Pipelines and Vessels
  - Part I   General
  - Part II   Installation
  - Part III   Resistance Heating Elements
  - Part IV   Impedance Heating
  - Part V   Induction Heating
  - Part VI   Skin Effect Heating
  - Part VII   Control and Protection

Motors

- 430 Motors, Motor Circuits and Controllers
  - Part I   General
  - Part II   Motor Circuit Conductors
  - Part III   Motor and Branch Circuit Overload Protection
  - Part IV   Motor Branch Circuit, Short Circuit and Ground Fault Protection

- Part V   Motor Feeder Short Circuit and Ground Fault Protection
- Part VI   Motor Control Circuits
- Part VII   Motor Controllers
- Part VIII   Motor Control Centers
- Part IX   Disconnecting Means
- Part X   Adjustable Speed Drive Systems
- Part XI   Over 600 Volts, Nominal
- Part XII   Protection of Live Parts, All Voltages
- Part XIII   Grounding, All Voltages
- Part XIV   Tables
- 440 Air Conditioning and Refrigerating Equipment
  - Part I   General
  - Part II   Disconnecting Means
  - Part III   Branch Circuit, Short Circuit and Ground Fault Protection
  - Part IV   Branch Circuit Conductors
  - Part V   Controllers for Motor Compressors
  - Part VI   Motor Compressor and Branch Circuit Overload Protection
  - Part VII   Provisions for Room Air Conditioners

Generation of Power for Utilization Equipment

Generators are appropriately located in Chapter 4, the USE chapter. In almost all electrical installations, the source of power to an occupancy will be supplied by an electrical utility and is called a *service.* When there is a need for an emergency or standby system, the most common or general backup system is an on-site standby generator (Figure 11–5).

Article 445 provides information and requirements for generators, which are equipment for general use. The *NEC®* addresses two additional energy systems that are considered "special" and are located in Chapter 6 in accordance with 90.3. Article 445 Generators, which addresses the generation requirements of Chapter 4, is not subdivided into separate parts.

Transformation of Systems for Utilization Equipment

Transformers are appropriately located in Chapter 4, the USE chapter. When utilization equipment operates at different system voltages in an electrical installation, transformers are used to derive new systems at the utilization voltage and system requirements (Figures 11–6a and b).

Phase converters, capacitors, resistors, and *reactors* are also located in Chapter 4. This equipment allows economical and efficient use of utilization equipment. For example, an older building or structure may have a 2-phase, 5-wire service. In order to use a 3-phase motor (much less expensive) in this occupancy, a phase converter would be required. In addition, power factor problems (wasting energy, extremely expensive) can be corrected through the application of capacitors.

DidYouKnow?

Article 445 provides information and requirements for generators. Generators are necessary to facilitate the use of electrical energy where an alternate source of supply is required.

FIGURE 11–5 Generators are necessary in many types of installations to supply power for utilization equipment.

Article 450 covers transformers, which are an essential part of most electrical installations providing an installation with the flexibility of deriving a new system to meet the requirements of utilization equipment. Phase converters, Article 455, allows for single-phase systems to derive a 3-phase system and older 2-phase electrical systems to utilize the 3-phase equipment, which is more common, cheaper, and readily available. While capacitors, resistors, and reactors do not transform or change a system, they are essential equipment, supplementing electrical systems in special applications.

FIGURE 11-6   Transformers are equipment for general use that provide separately derived systems for utilization equipment.

(a)

(b)

The four articles which address the transformation or adjustment of system requirements of Chapter 4 are as follows:

- 450 Transformers and Transformer Vaults (Including Secondary Ties)
  - Part I   General Provisions
  - Part II   Specific Provisions Applicable to Different Types of Transformers
  - Part III   Transformer Vaults
- 455 Phase Converters
  - Part I   General
  - Part II   Specific Provisions Applicable to Different Types of Phase Converters
- 460 Capacitors
  - Part I   600 Volts Nominal and Under
  - Part II   Over 600 Volts Nominal
- 470 Resistors and Reactors
  - Part I   600 Volts Nominal and Under
  - Part II   Over 600 Volts Nominal

DidYouKnow?

Artilce 450 contains information and requirements for transformers to facilitate the use of electrical energy. Where utilization equipment operates at a voltage level different from the supply, a transformer is used to derive a new system.

Storage of Power for Utilization Equipment

Batteries are appropriately located in Chapter 4, the USE chapter. There are many electrical installations that utilize batteries as a storage system to supply electrical energy to utilization equipment in the event of a power loss (Figure 11–7).

Storage batteries are used in many different electrical installations to provide power for utilization equipment in the event of a loss of power.

DidYouKnow?

Article 490 contains information and requirements for equipment rated at over 600 volts.

FIGURE 11-7 Batteries are equipment for general use that store energy for utilization equipment.

Article 480 provides information and requirements for the safe installation of battery-supplied systems. Article 480 Storage Batteries, which addresses the energy storage requirements of Chapter 4 is not subdivided into separate parts.

Equipment Over 600 Volts, Nominal

The specific requirements necessary for electrical equipment for general use, rated at over 600 volts nominal, are addressed by Article 490.

- 490 Equipment Over 600 Volts, Nominal
  - Part I   General
  - Part II   Equipment, Specific Provisions
  - Part III   Equipment, Metal Enclosed Power Switchgear and Industrial Control Assemblies
  - Part IV   Mobile and Portable Equipment
  - Part V   Electrode Type Boilers

KEY WORDS/CLUES FOR CHAPTER 4, "USE"

The control, connection, installation, protection, or construction specifications of general utilization equipment include:

- Lighting Fixtures/Luminaires
- Fixture Wire
- Cords and Cables
- Receptacles
- Cord Connectors, Attachment Plugs
- Switchboards
- Panelboards
- Industrial Control Panels
- Low-Voltage Lighting
- Appliances
- Fixed Electric Space Heating Equipment
- Fixed Outdoor Electric De-icing and Snow Melting Equipment
- Fixed Electric Heating Equipment for Pipelines and Vessels (Heat Trace)
- Motors, Motor Circuits and Controllers
- Air-Conditioning and Refrigerating Equipment
- Generators
- Transformers and Transformer Vaults
- Phase Converters
- Capacitors
- Resistors and Reactors
- Storage Batteries
- Electrical Equipment Over 600 Volts

SUMMARY

Chapter 4 in accordance with Section 90.3, applies generally to all electrical installations. Using the Codeology method, this chapter is given the nickname of the "Use" chapter, due to the scope covered. All of the material covered pertains to "electrical equipment for general use." The scope of this chapter is directed at equipment that *uses* electrical energy and associated equipment necessary for safe utilization.

The *NEC*® title for Chapter 4 is "Equipment for General Use." From this title the scope of this chapter can be described as "Information and

Rules on *Equipment for General Use* in Electrical Installations." Chapter 4 covers the entire electrical distribution system from the service point (connection to the utility) to the last receptacle or other outlet in the electrical system for electrical equipment, not wiring methods and wiring materials that are covered in Chapter 3.

See Figure 11–8 on page 200 for an example of how Chapter 4, along with Chapters 1, 2, and 3 build the foundation or backbone of all electrical installations.

REVIEW QUESTIONS

1. Does Chapter 4 of the *NEC*® apply to all electrical installations covered by the *NEC*®?

2. Chapter 4 is subdivided into how many articles?

3. The scope of Chapter 4 of the *NEC*® is dedicated to equipment for general use to facilitate the _____ of electrical energy.

4. What part of which article in Chapter 4 would address the installation of track lighting?

5. Chapter 4 of the *NEC*® addresses equipment for general use, which would include motors. What part of which article would apply if one were sizing motor overload protection?

6. Does 426.31 apply to resistance heating elements?

7. Which four articles in Chapter 4 cover requirements for the control of utilization equipment?

8. How many parts is Article 490 subdivided into?

9. Does Part III of Article 440 apply to the installation of portable room air conditioners?

10. Which article in Chapter 4 covers requirements for generation of power for utilization equipment?

FIGURE 11-8 Chapter 4 will apply generally in all electrical installations.

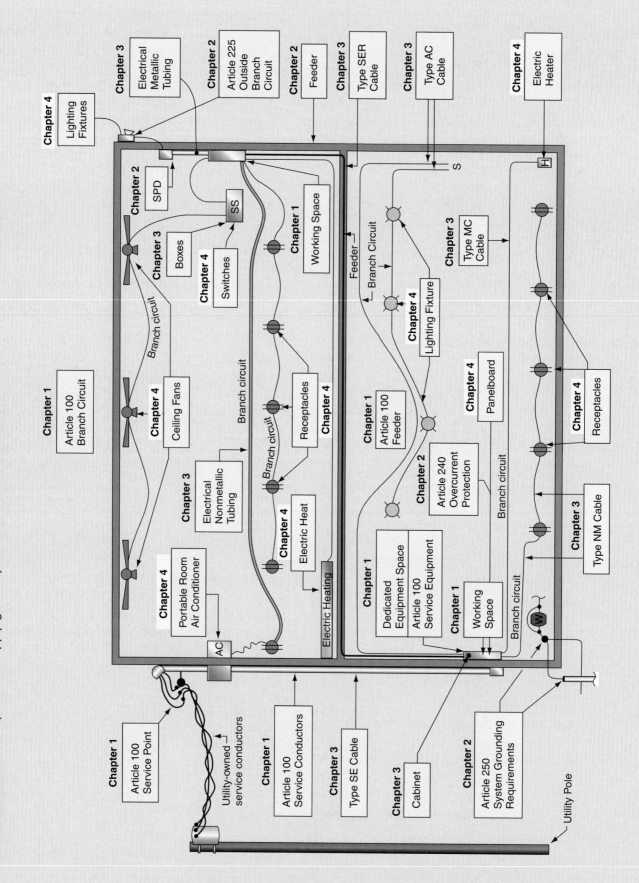

PRACTICE PROBLEMS

Using the Codeology Method

The following questions and steps to find the answer are designed to illustrate the Codeology method. While the steps to find your answer may seem lengthy, they are designed to illustrate the thought process to find the answer. The easiest way to make the Codeology method work is to silently "talk to yourself." Walk through these steps by silently talking to yourself and Codeology will be a natural response for quickly and accurately finding needed information in the *NEC®*.

Read the following statements and follow step by step using your codebook and the Codeology method.

1. The *NEC®* considers any lighting track identified for use exceeding _____ amps as "heavy duty lighting track."

Step #1

Qualify the question or need, look for key words
    This question is about track lighting.
    Key words: *"heavy duty track lighting"*
    Think equipment for general use, think USE

Step #2

Go to the table of contents, Chapter 4, the USE chapter
    Find the correct article   **410 Luminaires Lampholders and Lamps**

Step #3

Further qualify your question or need and get in the right part of the article
    Key words: *"ampacity rating for heavy duty track"*
    Look in **Part XV Lighting Track**

Step #4

Read each section title until the correct section is found
    Identify correct section or subdivision and find answer
    Correct section/subdivision
    Section   **410.153 Heavy-Duty Lighting Track**

2. In general, a disconnecting means is required for all motors. The disconnecting means shall be located _____ _____ _____ the motor location and the driven machinery location.

Step #1

Qualify the question or need, look for key words
    This question is about motors.
    Key words: *"disconnect location"*
    Think equipment for general use, think USE

Step #2

Go to the table of contents, Chapter 4, the USE chapter
    Find the correct article   **430 Motors, Motor Circuits and Controllers**

Step #3

Further qualify your question or need and get in the right part of the article
    Key words: *"motor, disconnect location"*
    Look in **Part IX Disconnecting Means**

Step #4

Read each section title until the correct section is found
    Identify correct section or subdivision and find answer
    Correct section/subdivision
    Section   **430.102 Location**
    First-Level Subdivision   **(B) Motor**

3. Twelve lead acid-type battery cells are connected in series. What is the total nominal voltage of these cells connected in series?

Step #1

Qualify the question or need, look for key words
    This question is about batteries.
    Key words: *"lead acid battery cells"*
    Think equipment for general use, think USE

Step #2

Go to the table of contents, Chapter 4, the USE chapter

Find the correct article   **480 Storage Batteries**

Step #3

Further qualify your question or need and get in the right part of the article

Key words: *"voltage of lead-acid battery cells"*

Article 480 is not separated into parts.

Step #4

Read each section title until the correct section is found

Identify correct section or subdivision and find answer

Correct section/subdivision

Section   **480.2 Definitions**

Nominal Battery Voltage

4. Each unit of fixed electric space-heating equipment shall be provided with a nameplate giving the identifying name and the normal rating in volts and watts or in volts and amperes. This nameplate shall be located so as to be visible or easily accessible after _____.

Step #1

Qualify the question or need, look for key words

This question is about fixed electric space-heating equipment.

Key words: *"fixed electric space-heating equipment"*

Think equipment for general use, think USE

Step #2

Go to the table of contents, Chapter 4, the USE chapter

Find the correct article   **424 Fixed Electric Space-Heating Equipment**

Step #3

Further qualify your question or need and get in the right part of the article

Key words: *"marking location for fixed electric space-heating equipment"*

Look in **Part IV Marking of Heating Equipment**

Step #4

Read each section title until the correct section is found

Identify correct section or subdivision and find answer

Correct section/subdivision

Section   **424.28 Nameplate**

First-Level Subdivision   **(B) Location**

5. A cord supplying a 240-volt room air conditioner shall not be longer than _____ feet.

Step #1

Qualify the question or need, look for key words

This question is about a room air conditioner.

Key words: *"room air conditioner"*

Think equipment for general use, think USE

Step #2

Go to the table of contents, Chapter 4, the USE chapter

Find the correct article   **440 Air-Conditioning and Refrigerating Equipment**

Step #3

Further qualify your question or need and get in the right part of the article

Key words: *"cord length for room air conditioner"*

Look in **Part VII Provisions for Room Air Conditioners**

Step #4

Read each section title until the correct section is found

Identify correct section or subdivision and find answer

Correct section/subdivision

Section   **440.64 Supply Cords**

6. Type SJOW flexible cord is limited to installations of _____ volts or less.

Step #1

Qualify the question or need, look for key words

This question is about a type of flexible cord.

Key words: *"type SJOW flexible cord"*

Think equipment for general use, think USE

Correct section/subdivision

Section   **408.3 Support and Arrangement of Busbars and Conductors**

First-Level Subdivision   **(E) Phase Arrangement**

Step #2

Go to the table of contents, Chapter 4, the USE chapter

Find the correct article   **400 Flexible Cords and Cables**

8. Receptacles incorporating an isolated grounding connection intended for the reduction of electrical noise (electromagnetic interference) shall be identified by an _____ _____ located on the face of the receptacle.

Step #3

Further qualify your question or need and get in the right part of the article

Key words: *"voltage limitations for type SJOW flexible cord"*

Look in **Part I General**

Step #1

Qualify the question or need, look for key words

This question is about a receptacle.

Key words: *"isolated grounding type receptacle"*

Think equipment for general use, think USE

Step #4

Read each section title until the correct section is found

Identify correct section or subdivision and find answer

Correct section/subdivision

Section   **400.4 Types**

Table 400.4 Flexible Cords and Cables

Step #2

Go to the table of contents, Chapter 4, the USE chapter

Find the correct article   **406 Receptacles, Cord Connectors, and Attachment Plugs (Caps)**

7. The phase arrangement in a panelboard or switchboard is required to be _____ when viewed from the front to back, top to bottom, or left to right as viewed from the front of the panel-board or switchboard.

Step #3

Further qualify your question or need and get in the right part of the article

Key words: *"identification of an isolated grounding type receptacle"*

Article 406 is not separated into parts.

Step #1

Qualify the question or need, look for key words

This question is about a panelboard.

Key words: *"panelboards ans switchboards"*

Think equipment for general use, think USE

Step #4

Read each section title until the correct section is found

Identify correct section or subdivision and find answer

Correct section/subdivision

Section   **406.2 Receptacle Rating and Type**

First-Level Subdivision   **(D) Isolated Ground Receptacles**

Step #2

Go to the table of contents, Chapter 4, the USE chapter

Find the correct article   **408 Switchboards and Panelboards**

Step #3

Further qualify your question or need and get in the right part of the article

Key words: *"phase arrangement"*

Look in **Part I General**

Step #4

Read each section title until the correct section is found

Identify correct section or subdivision and find answer

9. All types of switches with a marked "OFF" position shall be constructed to completely disconnect all _____ conductors to the load being controlled.

Step #1

Qualify the question or need, look for key words
　This question is about switches.

　　Key words:　*"switches with a marked "OFF" position"*

　Think equipment for general use, think USE

Step #2

Go to the table of contents, Chapter 4, the USE chapter
　Find the correct article　　**404 Switches**

Step #3

　Further qualify your question or need and get in the right part of the article
　　Key words:　*"construction of switches with a marked "OFF" position"*

　Look in **Part II Construction Specifications**

Step #4

Read each section title until the correct section is found
　Identify correct section or subdivision and find answer
　Correct section/subdivision
　Section　　**404.15 Marking**
　First-Level Subdivision　　**(B) Off Indication**

10. Capacitors are to be provided with a means to discharge stored energy. For capacitors rated 600 volts and less, the residual voltage shall be reduced to 50 volts or less within _____ after the capacitor is disconnected.

Step #1

Qualify the question or need, look for key words
　This question is about capacitors.

　　Key words:　*"a means to discharge stored energy"*

　Think equipment for general use, think USE

Step #2

Go to the table of contents, Chapter 4, the USE chapter
　Find the correct article　　**460 Capacitors**

Step #3

Further qualify your question or need and get in the right part of the article
　　Key words:　*"600 volts and less, a means to discharge stored energy, voltage"*

　Look in **Part I 600 Volts, Nominal, and Under**

Step #4

Read each section title until the correct section is found
　Identify correct section or subdivision and find answer
　Correct section/subdivision
　Section　　**460.6 Discharge of Stored Energy**
　First-Level Subdivision　　**(A) Time of Discharge**

11. Lighting systems operating at 30 volts or less shall be supplied from a maximum of a _____ amp branch circuit.

Step #1

Qualify the question or need, look for key words
　This question is about lighting systems operating at 30 volts or less.

　　Key words:　*"lighting systems operating at 30 volts or less"*

　Think equipment for general use, think USE

Step #2

Go to the table of contents, Chapter 4, the USE chapter
　Find the correct article　　**411 Lighting Systems Operating at 30 Volts or Less**

Step #3

Further qualify your question or need and get in the right part of the article
　　Key words:　*"maximum size branch circuit for lighting systems operating at 30 volts or less"*

　Article 411 is not separated into parts.

Step #4

Read each section title until the correct section is found
　Identify correct section or subdivision and find answer
　Correct section/subdivision
　Section　　**411.6 Branch Circuit**

12. A fixed storage-type water heater that has a capacity of _____ gallons or less shall be considered a continuous load for the purposes of sizing branch circuits.

Step #1

Qualify the question or need, look for key words
This question is about a water heater.

Key words: *"fixed storage-type water heater"*

Think equipment for general use, think USE

Step #2

Go to the table of contents, Chapter 4, the USE chapter
Find the correct article   **422 Appliances**

Step #3

Further qualify your question or need and get in the right part of the article
Key words: *"size of fixed storage-type water heater"*

Look in **Part II Installation**

Step #4

Read each section title until the correct section is found
Identify correct section or subdivision and find answer
Correct section/subdivision
Section   **422.13 Storage-Type Water Heaters**

13. Thermal insulation shall not be installed above a recessed luminaire or within _____ inches of the recessed luminaire's enclosure, wiring compartment, or ballast unless it is identified for contact with insulation, as type IC.

Step #1

Qualify the question or need, look for key words
This question is about a lighting fixture or luminaire.

Key words: *"recessed luminaire"*

Think equipment for general use, think USE

Step #2

Go to the table of contents, Chapter 4, the USE chapter
Find the correct article   **410 Luminaires, Lampholders and Lamps**

Step #3

Further qualify your question or need and get in the right part of the article
Key words: *"thermal insulation near a recessed luminaire"*

Look in **Part XI Special Provisions for Flush and Recessed Luminaires**

Step #4

Read each section title until the correct section is found
Identify correct section or subdivision and find answer
Correct section/subdivision
Section   **410.116 Clearance and Installation**
First-Level Subdivision   **(B) Installation**

14. In general, motor circuit conductors supplying a single motor in a continuous duty application shall be sized at _____% of the motor full load current.

Step #1

Qualify the question or need, look for key words
This question is about a motor installation.

Key words: *"motor circuit conductors"*

Think equipment for general use, think USE

Step #2

Go to the table of contents, Chapter 4, the USE chapter
Find the correct article   **430 Motors, Motor Circuits and Controllers**

Step #3

Further qualify your question or need and get in the right part of the article
Key words: *"motor circuit conductors, sizing for single motor, continuous duty"*

Look in **Part II Motor Circuit Conductors**

Step #4

Read each section title until the correct section is found
Identify correct section or subdivision and find answer
Correct section/subdivision
Section   **430.22 Single Motor**
First-Level Subdivision   **(A) General**

15. When single-phase loads are connected on the load side of a phase converter, they shall not be connected to the _____ phase.

Qualify the question or need, look for key words

This question is about a phase converter.

Key words: *"phase converter"*

Think equipment for general use, think USE

Go to the table of contents, Chapter 4, the USE chapter

Find the correct article   **455 Phase Converters**

Further qualify your question or need and get in the right part of the article

Key words: *"single-phase loads on a phase converter"*

Look in **Part I General**

Read each section title until the correct section is found

Identify correct section or subdivision and find answer

Correct section/subdivision

Section   **455.9 Connection of Single-Phase Loads**

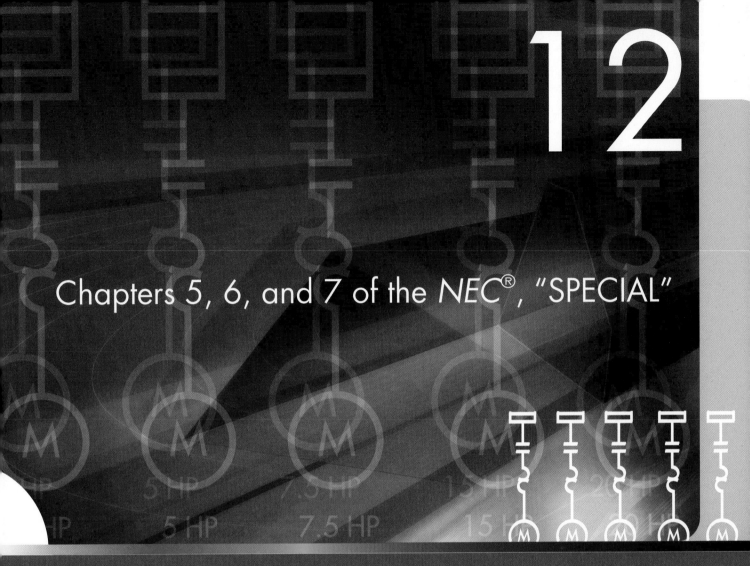

12

Chapters 5, 6, and 7 of the *NEC*®, "SPECIAL"

O U T L I N E

OBJECTIVES

After completing this unit, you should be able to:
1. Associate the Codeology title for *NEC*® Chapter 5 as "Special Occupancies"
2. Associate the Codeology title for *NEC*® Chapter 6 as "Special Equipment"
3. Associate the Codeology title for *NEC*® Chapter 7 as "Special Conditions"
4. Identify the special type of information and requirements contained in Chapters 5, 6, and 7 for supplementing and/or modifying the requirements of Chapters 1 through 4
5. Recognize key words and clues for locating answers in *NEC*® Chapters 5, 6, and 7, the Special chapters
6. Identify exact sections, subdivisions, list items, etc. to justify answers for all questions referring to Chapter 5, 6, and 7 articles
7. Recognize that Chapters 5, 6, and 7 numbering are the 500-, 600-, and 700-series
8. Recognize, recall, and become familiar with articles contained in Chapters 5, 6, and 7

OVERVIEW

In accordance with 90.3, Chapters 1 through 4 are general in scope and provide installation requirements for the entire electrical system. Chapters 5, 6, and 7 deal with special requirements and supplement or modify the first four chapters. All of the requirements in these Special chapters are modifications of basic rules or supplemental requirements to address special needs. There are 62 articles in the Special chapters. It is absolutely necessary for the user of this Code to become familiar with the different types of special occupancies, equipment, and conditions.

ARRANGEMENT OF THE SPECIAL CHAPTERS

The need for modifications and supplemental requirements of the Special chapters must be identified before any work is started on an electrical installation. Each stage of an electrical installation will be affected when a special situation is encountered. The special requirements of Chapters 5, 6, and 7 must be considered in each step of the installation to prevent serious misapplication of the *NEC*®.

For example, a new hospital is to be constructed and the electrical installation is being considered. A hospital is considered a special occupancy because of the special needs for uninterrupted power, protection of patients, and many other special considerations. The *NEC*® covers the special needs of a hospital in Article 517 Health Care Facilities.

THE GENERAL, PLAN, BUILD, AND USE STAGES ARE MODIFIED AND SUPPLEMENTED BY CHAPTERS 5, 6, AND 7

For example, Chapter 5 Special Occupancies will affect the PLAN stage of the hospital. An example of how Article 517 will modify and supplement the PLAN stage of the electrical installation is special requirements for grounding, branch circuits, and receptacle locations for patient care areas. These special requirements will require specific wiring methods to the patient care areas thereby having an effect on the BUILD stage of this installation. The USE stage of this installation is modified and supplemented in Chapter 6 Special Equipment, including X-ray equipment. All stages of the installation will be further modified and supplemented by Chapter 7 Special Conditions when emergency systems are installed to meet life safety requirements.

NEC® CHAPTER 5 "SPECIAL OCCUPANCIES"

The Codeology title for Chapter 5 is "Special Occupancies." All stages of an electrical installation will be modified or supplemented when special occupancies are involved. There are 28 articles in Chapter 5 and these can be separated into the following groupings:

Hazardous Locations

Article 500 Hazardous (Classified) Locations, Classes I, II, and III, Divisions 1 and 2

Article 501 Class I Locations

Article 502 Class II Locations

Article 503 Class III Locations

Article 504 Intrinsically Safe Systems

Article 505 Class I, Zone 0, 1, and 2 Locations

Article 506 Zone 20, 21, and 22 Locations for Combustible Dusts, Fibers, and Flyings

Article 510 Hazardous (Classified) Locations, Specific

Article 511 Commercial Garages, Repair and Storage

DidYouKnow?

The *NEC*® provides the Code user with occupancy-specific requirements throughout the general Chapters 1 through 4. Where an occupancy requires a modification or supplemental requirements, it is considered "Special" and is located in Chapter 5.

HAZARDOUS LOCATIONS

Articles 500 through 516 contain modifications and supplemental requirements for occupancies that contain, process, manufacture, or store materials that could cause a fire or explosion due to flammable gases or vapors, flammable liquids, combustible dust/s or ignitable fibers and flyings (Figures 12–1a and b).

Article 500 sets the stage for the application of Articles 501, 502, 503, and 504. Section 500.2 contains definitions that apply to all "Hazardous Location" articles from 500 through 516 except for the two "Zone Classification" articles, 505 and 506, which provide an alternative to the Class I, II, and III systems. Article 500 provides the basic information and requirements for the application of the Class I, II, and III systems for hazardous locations. The following key provisions of article 500 apply to all hazardous location articles except for the Zone System Articles 505 and 506.

DidYouKnow?

Hazardous locations require wiring methods and materials to prevent fires and/or explosions. These "Special" requirements are logically located in Chapter 5.

FIGURE 12–1 Hazardous locations are considered special occupancies. These include all areas that contain, use, or manufacture materials that are easily ignitable, produce volatile vapors or are explosive under given conditions.

(a)

(b)

DidYouKnow?

Article 500 provides general information and requirements necessary for the application of Articles 501, 502, 503, and 504.

Article 500 Hazardous (Classified) Locations, Classes I, II, and III, Divisions 1 and 2

500.2 Definitions

The following definitions apply to Articles 501, 502, 503, 504, 510, 511, 513, 514, 515, and 516:

Associated Nonincendive Field-Wiring Apparatus

Combustible Gas Detection System

Control Drawing

Dust Ignitionproof

Dusttight

Electrical and Electronic Equipment

Explosionproof Apparatus

Hermetically Sealed

Nonincendive Circuit

Nonincendive Component

Nonincendive Equipment

Nonincendive Field Wiring

Nonincendive Field Wiring-Apparatus

Oil Immersion

Purged and Pressurized

Unclassified Locations

500.4 General

500.4(A) Documentation This requires that all areas designated as hazardous locations are properly documented and the documentation be available to designers, installers, maintainers, and operators at the location.

500.4(B) Reference Standards These five FPNs provide information to the Code user by referencing other applicable codes to assist in proper application of the *NEC®* requirements.

500.5 Classification of Locations

500.5(A) Classification of Locations Hazardous locations are classified depending on the properties of the flammable vapors, liquids, gases, or flammable liquid-produced combustible dusts or fibers that may be present, and the likelihood that a flammable or combustible concentration or quantity is present.

500.5(B) Class I Locations Class I locations are those in which *flammable gases* or *flammable liquid-produced vapors* are or may be present in the air in quantities sufficient to produce explosive or ignitable mixtures (Figure 12–2).

Class I Division 1

- Ignitable concentrations exist under normal operations.
- Ignitable concentrations may exist due to repair, maintenance, or leaks
- Ignitable concentrations exist due to processes, breakdown, or faulty equipment.

Class I Division 2

- Locations in which volatiles are handled, processed, or used in closed containers.
- Locations in which positive ventilation prevents accumulation of gases/vapors.
- Areas adjacent to Class I Division 1 locations.

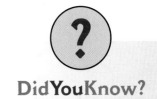

Did You Know?

Hazardous locations may be classified as Class I, II or, III; or the Zone System may be applied.

FIGURE 12-2

When flammable gases or vapors are present, the occupancy is considered a Class 1 Hazardous Location.

500.5(C) Class II Locations Class II locations are those that are hazardous because of the presence of combustible dust (Figure 12–3).

Class II Division 1

- Ignitable concentrations of combustible dust exist under normal operations.

- Where mechanical or machinery failure, repair, maintenance or leaks could create ignitable concentrations of dust.

- Locations in which metal dusts exist, including, but not limited to, aluminum and magnesium.

FIGURE 12–3 The storage of grain can produce a hazardous location due to the presence of combustible dust.

Class II Division 2

- Ignitable concentrations of combustible dust may exist due to abnormal operations.

- Locations in which combustible dust is present but not in sufficient amounts unless a malfunction of equipment or process occurs.

- Locations in which accumulating dust interferes with heat dissipation and/or could be ignited through equipment failure.

500.5(D) Class III Locations Class III locations are those that are hazardous because of the presence of easily ignitible fibers or flyings, but in which such fibers or flyings are not likely to be in suspension in the air in quantities sufficient to produce ignitable mixtures.

Class III Division 1

- Locations in which easily ignitable fibers or materials producing combustible flyings are handled, manufactured, or used.

  Class III Division 2

- Locations in which easily ignitable fibers are stored or handled other than in the process of manufacture.

500.6 Material Groups

This section contains "group classifications" for combustible gases, vapors, and dusts. These group classifications are as follows:

- Class I Groups A, B, C, and D
- Class II Groups E, F, and G

500.7 Protection Techniques

The permitted protection techniques for electrical and electronic equipment are listed in 11 first-level subdivisions.

500.8 Equipment

Minimum equipment requirements are listed, including:

- Suitability
- Approval
- Marking
- Temperature
- Threading

Article 501 Class I Locations

Part I General

- Scope
- General Rules

Part II Wiring

- Wiring Methods
- Sealing and Drainage
- Conductor Insulation
- Uninsulated Exposed Parts
- Grounding and Bonding
- Surge Protection

Part III Equipment

- Transformers and Capacitors
- Meters, Instruments, and Relays
- Switches, Circuit Breakers, Motor Controllers and Fuses
- Control Transformers and Resistors
- Motors and Generators

DidYouKnow?

The application of equipment in a hazardous location requires that the proper group classification be chosen with respect to the gas, vapors, dusts, fibers, or flyings present.

DidYouKnow?

Article 501 provides information and requirements for Class I Division 1 and 2 locations where fire or explosion hazards exist due to flammable gases or vapors or flammable liquids.

- Luminaires
- Utilization Equipment
- Flexible Cords
- Receptacles and Attachment Plugs
- Signaling, Alarm, Remote Control, and Communications Systems

Article 502 Class II Locations

Part I General

- Scope
- General Rules

Part II Wiring

- Wiring Methods
- Sealing
- Uninsulated Exposed Parts
- Grounding and Bonding
- Surge Protection
- Multiwire Branch Circuits

Part III Equipment

- Transformers and Capacitors
- Ventilating Piping
- Switches, Circuit Breakers, Motor Controllers and Fuses
- Control Transformers and Resistors
- Motors and Generators
- Luminaires
- Utilization Equipment
- Flexible Cords
- Receptacles and Attachment Plugs
- Signaling, Alarm, Remote Control, and Communications Systems
- Meters Instruments and Relays

Article 503 Class III Locations

Part I General

- Scope
- General Rules

Part II Wiring

- Wiring Methods
- Uninsulated Exposed Parts
- Grounding and Bonding

Part III Equipment

- Transformers and Capacitors
- Ventilating Piping
- Switches, Circuit Breakers, Motor Controllers and Fuses
- Control Transformers and Resistors
- Motors and Generators
- Luminaires
- Utilization Equipment

DidYouKnow?

Article 502 provides information and requirements for Class II, Division 1 and 2 locations where fire or explosion hazards may exist due to combustible dust.

DidYouKnow?

Article 503 provides information and requirements for Class III Division 1 and 2 locations where fire or explosion hazards may exist due to ignitable fibers or flyings.

- Flexible Cords
- Receptacles and Attachment Plugs
- Signaling, Alarm, Remote Control, and Local Loudspeaker Intercommunications Systems
- Electric Cranes, Hoists, and Similar Equipment
- Storage Battery Charging Equipment

Article 504 Intrinsically Safe Systems

An intrinsically safe circuit is defined in 504.2 as a circuit in which "any spark or thermal effect is incapable of causing ignition of a mixture of flammable or combustible material in air under prescribed test conditions." Intrinsically safe systems are by design incapable of causing fire or explosion. These systems are designed and identified for use in hazardous locations.

ZONE CLASSIFICATION SYSTEM

The Zone Classification System provides the user of this Code with an alternate method to the Class I, II, and III systems outlined in Articles 500, 501, 502, and 503. Article 505 covers Class I locations and Article 506 covers Class II and III locations.

Article 505 Class I Zone 0, 1, and 2 Locations

Class I Zone 0
- Ignitable concentrations are present continuously.
- Ignitable concentrations are present for long periods of time.

Class I Zone 1
- Ignitable concentrations exist under normal operations.
- Ignitable concentrations may exist due to repair, maintenance, or leaks
- Ignitable concentrations exist due to processes, breakdown, or faulty equipment.
- Areas adjacent to Class I Zone 0 locations.

Class I Zone 2
- Locations in which ignitable concentrations are not likely and could only occur briefly.
- Locations in which volatiles are handled, processed, or used in closed containers.
- Locations in which positive ventilation prevents accumulation of gases/vapors.
- Areas adjacent to Class I Zone 1 locations.

Article 506 Zone 20, 21, and 22 Locations for Combustible Dusts or Ignitable Fibers or Flyings

Zone 20
- Ignitable concentrations of combustible dust or ignitable fibers or flyings are present continuously.

DidYouKnow?

The Zone Classification System provides the Code user with an alternate method to the Class I, II, and III Systems.

- Ignitable concentrations of combustible dust or ignitable fibers or flyings are present for long periods of time.

Zone 21

- Ignitable concentrations of combustible dust or ignitable fibers or flyings exist under normal operations.
- Ignitable concentrations of combustible dust or ignitable fibers or flyings may exist due to repair, maintenance, or leaks.
- Ignitable concentrations of combustible dust or ignitable fibers or flyings exist due to processes, breakdown, or faulty equipment.
- Areas adjacent to Zone 20 locations.

Zone 22

- Locations in ignitable concentrations of combustible dust or ignitable fibers or flyings are not likely and could only occur briefly.
- Locations in which combustible dust or ignitable fibers or flyings are handled, processed, or used in closed containers.
- Areas adjacent to Zone 21 locations.

Specific Class 1, 2, and 3 Locations

Article 510 sets the stage for the application of Articles 511, 513, 514, 515, 516, and 517. Note that Article 510 also references Article 517 for health care facilities. This reference is due to the possibility of explosive gases or vapors that may be used in inhalation anesthetizing locations. Article 517 is not grouped in this textbook with the specific Class I, II, and III locations due to the many other requirements that make Article 517 special.

Article 510 clearly states that the provisions of Articles 500, 501, 502, and 504 will apply to the specific locations that follow unless modified or supplemented by Articles 511, 513, 514, 515, 516, and 517.

Article 510 Hazardous (Classified) Locations, Specific

This article clearly points out that the following "specific hazardous location" articles will modify or supplement Articles 500 through 504. The following articles cover occupancies that are or may be hazardous and the rules contained within will modify or supplement Articles 500 through 504.

Article 511 Commercial Garages, Repair and Storage

Includes service and repair operations for cars, buses, trucks, and tractors when volatile, flammable liquids or gases are used for fuel or power (Figure 12–4).

Article 513 Aircraft Hangars

Includes buildings or structures containing aircraft with volatile flammable liquids.

Article 514 Motor Fuel Dispensing Facilities

Includes gas stations, marine/motor fuel dispensing, motor fuel dispensing indoors and outdoors (Figure 12–5).

DidYouKnow?

Article 510 exists only to inform the Code user that the requirements of Articles 511 through 517 will supplement or modify Articles 500 through 504.

FIGURE 12-4 Commercial garages are hazardous locations due to the presence of volatile liquids and/or gases as a fuel source for vehicles.

FIGURE 12-5

Gas stations and any other location in which fuel is dispensed represent hazardous locations.

Article 515 Bulk Storage Plants

Includes property where flammable liquids are received by tank vessel, pipelines, tank car, or tank vehicle and are stored or blended in bulk for the purpose of distributing such liquids by tank vessel, pipeline, tank car, tank vehicle, portable tank, or container.

Article 516 Spray Application, Dipping and Coating Processes

Includes locations in which regular or frequent application of flammable or combustible liquids and powders occurs by spray operations.

HEALTH CARE FACILITIES

Health care facilities are special occupancies that require the modification of the general rules in Chapters 1 through 4 of the *NEC*®. Health care facili-

DidYouKnow?

Chapters 1 through 4 of the *NEC*® are supplemented and modified by Article 517 to address the specific requirements of Health Care Facilities.

ties are buildings or portions of buildings in which medical, dental, psychiatric, nursing, obstetrical, or surgical care are provided (Figure 12–6). Health care facilities include but are not limited to hospitals, nursing homes, limited care facilities, clinics, medical and dental offices, and ambulatory care centers, whether permanent or movable.

FIGURE 12–6 An emergency room of a hospital represents a special occupancy. The supplemental information and rules of Article 517 address the unique requirements for health care facilities.

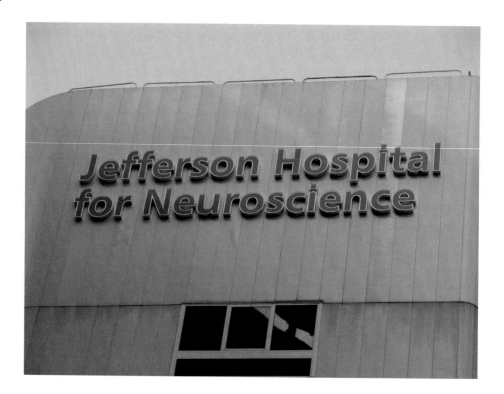

There are 39 definitions in Section 517.2 to deal with the special requirements of a health care facility. There are many different types of locations covered in Article 517. The definition of *health care facilities* in 517.2 outlines locations covered by this article. Article 517 is subdivided into parts as follows:

Article 517 Health Care Facilities

Part I	General
Part II	Wiring and Protection
Part III	Essential Electrical System
Part IV	Inhalation Anesthetizing Locations
Part V	X-Ray Installations
Part VI	Communications, Signaling Systems, Data Systems, Fire Alarm Systems, and Systems Less Than 120 Volts, Nominal
Part VII	Isolated Power Systems

ASSEMBLY OCCUPANCIES FOR 100 OR MORE PERSONS

Special consideration of electrical installations is provided by the *NEC*® to all locations where large numbers of people are intended to assemble (Figure 12–7). Article 518 Assembly Occupancies, contains modifications and supplemental requirements for all assembly locations. As per 518.1, this article is intended to cover all buildings or portions of buildings or structures intended for groups of 100 or more persons for meetings, worship, entertainment, eating, drinking, amusement, awaiting transportation, or any other similar purpose. Examples of assembly occupancies include but are not limited to the following:

FIGURE 12–7 Assembly occupancies include all rooms or areas in venues intended for the assembly of 100 or more people.

Article 518 Assembly Occupancies

Armories
Assembly halls
Auditoriums
Bowling lanes
Club rooms
Conference rooms
Courtrooms

Dance halls
Dining/Drinking facilities

Exhibition halls
Gymnasiums
Mortuary chapels
Multipurpose rooms
Museums
Places of awaiting transportation
Places of religious worship
Pool rooms
Restaurants
Skating rinks

DidYouKnow?

Article 518 requires the use of only wiring methods recognized in 518.4 for all buildings or portions of buildings intended for the gathering of 100 or more persons.

ENTERTAINMENT VENUES

Entertainment venues have many special electrical installation requirements. Theatres and motion picture venues have special needs for lighting and equipment and are designed to be attended in many cases by 100 or more persons, making them assembly occupancies. These locations are also intended to be occupied in low-light-level situations. Circuses, carnivals, and fairs also have very special needs due to being outdoors in all weather with large numbers of people in attendance.

Article 520 Theatres, Audience Areas of Motion Picture and Television Studios, Performance Areas, and Similar Locations

Article 522 Control Systems for Permanent Amusement Attractions

Article 525 Carnivals, Circuses, Fairs, and Similar Events

Article 530 Motion Picture and Television Studios and Similar Locations

Article 540 Motion Picture Projection Rooms

AGRICULTURAL BUILDINGS FOR POULTRY, LIVESTOCK, AND FISH

Agricultural buildings containing poultry, livestock, or fish present very special needs for electrical installations. An example is the requirement of 547.10 for equipotential planes. This bonding requirement is necessary to prevent livestock from an electrical shock.

Article 547 Agricultural Buildings

Where excessive dust and dust with water may accumulate in agricultural buildings and all agricultural buildings used for poultry, livestock, and fish confinement systems, special electrical installation requirements are necessary.

MANUFACTURED BUILDINGS, DWELLINGS, AND RECREATIONAL STRUCTURES

When buildings and/or structures are manufactured and are then moved into place for use by persons for occupancies including storage, places of employment, dwelling units, and recreational uses, special electrical construction and installation requirements exist.

Article 545 Manufactured Buildings

Article 550 Mobile Homes, Manufactured Homes, and Mobile Home Parks

Article 551 Recreational Vehicles and Recreational Vehicle Parks

Article 552 Park Trailers

STRUCTURES ON OR ADJACENT TO BODIES OF WATER

Electrical installations for buildings and structures designed for use floating on, above, or adjacent to bodies of water require special electrical installation considerations (Figure 12–8). Articles 553 and 555 address the special

FIGURE 12-8 As special occupancies, marinas have specific needs for their electrical systems due to their close proximity to water and boats.

DidYouKnow?

Temporary Installations are included in Chapter 5 because as a structure or building is constructed (everything from a dwelling unit to a health care facility), it is supplied by a temporary system and is considered "Special."

electrical installation needs of occupancies designed to be floating on, above, or adjacent to bodies of water.

Article 553 Floating Buildings

Article 555 Marinas and Boatyards

TEMPORARY INSTALLATIONS

Article 590 Temporary Installations

Temporary power when used for construction, holiday decorative lighting, emergencies, and tests requires special installation considerations (Figure 12–9).

FIGURE 12–9   Article 590 addresses the requirements for a temporary electrical installation.

Key provisions of Article 590 Temporary Installations include:

590.3 Time Constraints

(A) **During the Period of Construction.** Temporary wiring is permitted for the length of the project.
(B) **90 Days.** Holiday and similar lighting is not permitted to exceed 90 days.
(C) **Emergencies and Tests.** Temporary installations are permitted for emergencies and tests.
(D) **Removal.** Temporary installations must be removed immediately after use.

590.6 Ground-Fault Protection for Personnel

(A) **Receptacle Outlets.** All 125-volt, single-phase, 15-, 20-, and 30-ampere receptacle outlets that are in use by personnel must be GFCI protected. When a receptacle is part of the permanent wiring of the building or structure and is to be used by persons, GFCI protection listed as "portable" must be provided.

NEC® CHAPTER 6 "SPECIAL EQUIPMENT"

The Codeology title for Chapter 6 is "Special Equipment." Chapter 4 of the *NEC®* is titled "Equipment for General Use" and covers the basics. Chapter 6 covers "Special Equipment" and modifies or supplements the first four chapters to meet the special needs of the equipment covered in this chapter as follows:

Article 600 Electric Signs and Outline Lighting

This article covers all electric sign and outline lighting installations, including neon tubing (Figure 12–10).

FIGURE 12–10 Electric signs are special equipment and are covered in Chapter 6 of the *NEC®*.

Article 604 Manufactured Wiring Systems

This article covers field-installed wiring, using manufactured subassemblies.

Article 605 Office Furnishings (Consisting of Lighting Accessories and Wired Partitions)

This article covers wiring and lighting equipment on re-locatable wired partitions.

Article 610 Cranes and Hoists

All electrical equipment and wiring used with cranes, monorail hoists, hoists, and all runways is addressed in this article.

Article 620 Elevators, Dumbwaiters, Escalators, Moving Walks, Wheelchair Lifts, and Stairway Chair Lifts

The installation of electrical equipment and wiring for elevators, dumbwaiters, escalators, moving walks, wheelchair lifts, and stairway chair lifts are covered by this article.

Article 625 Electric Vehicle Charging Systems

Conductors and equipment external to an electric vehicle related to charging the electric vehicle is covered by this article.

Article 626 Electrified Truck Parking Spaces

This article covers electrical conductors, devices and equipment that connect trucks (tractor-trailer rigs) and associated refrigeration units to a source of electrical energy.

Article 630 Electric Welders

This article covers the equipment and installation necessary for electric arc welding, resistance welding, plasma cutting, and other welding/cutting methods.

Article 640 Audio Signal Processing, Amplification, and Reproduction Equipment

This article covers equipment and wiring for audio signal generation, recording, processing, amplification, and reproduction; distribution of sound; public address; speech input systems; temporary audio system installations; and electronic organs or other electronic musical instruments.

Article 645 Information Technology Equipment

This article covers information technology equipment and systems in an information technology room only where the conditions of 645.4 are met.

Article 647 Sensitive Electronic Equipment

Installation and wiring of separately derived systems operating at 120 volts line-to-line and 60 volts to ground are covered by this article.

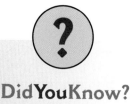

DidYouKnow?

The modifications and supplemental requirements of Article 645 are optional. This article only applies when the installer meets the requirements of 645.4.

Article 650 Pipe Organs

Electrical circuits and parts of pipe organs providing control of keyboards and sound are covered by this article.

Article 660 X-Ray Equipment

All X-ray equipment operating at any frequency or voltage is covered by this article.

Article 665 Induction and Dielectric Heating Equipment

Industrial and scientific applications of dielectric heating, induction heating, induction melting, and induction welding are covered by this article.

Article 668 Electrolytic Cells

This article covers the use of electrolytic cells for the production of aluminum, cadmium, chlorine, copper, fluorine, hydrogen peroxide, magnesium, sodium, sodium chlorate, and zinc.

Article 669 Electroplating

Installation of electrical components and equipment for electroplating, anodizing, electropolishing, and electrostripping are covered by this article.

Article 670 Industrial Machinery

Overcurrent protection for and the size of supply conductors and requirements for nameplate data of industrial machinery are covered by this article.

Article 675 Electrically Driven or Controlled Irrigation Machines

Electrically driven or controlled irrigation machines and branch circuits and controllers involved are covered by this article.

Article 680 Swimming Pools, Fountains, and Similar Installations

This article covers the construction and installation of electrical wiring for and equipment in or adjacent to all swimming, wading, therapeutic, and decorative pools, fountains, hot tubs, spas, and hydromassage bathtubs, whether permanently installed or storable, and to metallic auxiliary equipment, such as pumps, filters, and similar equipment (Figure 12–11).

Article 682 Natural and Artificially Made Bodies of Water

This article applies to all electrical installations made in or adjacent to natural or artificially made bodies of water not covered by other articles in the *NEC*®.

Article 685 Integrated Electrical Systems

This article covers integrated electrical systems when an orderly shutdown is necessary to ensure safe operation.

Article 690 Solar Photovoltaic Systems

Solar photovoltaic electrical energy systems including the array circuits, inverters, and controllers are covered by this article (Figure 12–12).

DidYouKnow?

680.2 contains 22 definitions that are essential for the proper application of these special requirements for pools, fountains, and similar installations.

FIGURE 12–11 Pools and spas are special equipment and are covered in Chapter 6 of the *NEC*®.

FIGURE 12–12 Solar photovoltaic systems are special equipment addressed in Chapter 6.

Article 692 Fuel Cell Systems

Stand-alone or interactive fuel cell power systems are covered by this article (Figure 12–13).

FIGURE 12–13 Fuel cell systems are special equipment addressed in Chapter 6.

Article 695 Fire Pumps

This article covers the installation of power sources, interconnecting circuits, and switching and control equipment for fire pumps. Note that this article does not apply to the pressure maintenance or jockey pumps.

NEC® CHAPTER 7 "SPECIAL CONDITIONS"

The Codeology title for Chapter 7 is "Special Conditions." Special conditions are required to meet the special needs of different types of occupancies and equipment. For example, when an alternate power source is required for emergency or legally required standby, it must be installed according to Chapters 1 through 4 and modified and/or supplemented in accordance with the requirements of Chapter 7 "Special Conditions."

Article 700 Emergency Systems

This article applies to the installation, operation, and maintenance of emergency systems consisting of circuits and equipment intended to supply, distribute, and control electricity for illumination, power, or both (Figure 12–14).

DidYouKnow?

Where an alternate power source is required or is desired, a "standby generator" is typically employed.

FIGURE 12–14 Emergency systems, which in many installations use a standby generator as the emergency source, represent special conditions and are addressed in Chapter 7.

Article 701 Legally Required Standby Systems

This article applies to the installation, operation, and maintenance of legally required standby systems consisting of circuits and equipment intended to supply, distribute, and control electricity for illumination, power, or both.

Article 702 Optional Standby Systems

This article covers the installation and operation of optional standby systems.

Article 705 Interconnected Electric Power Production Sources

Electric power production sources that operate in parallel with a primary source are covered by this article.

Article 708 Critical Operations Power Systems (COPS)

This article will apply to designated critical opertion areas when required by municipal, state, federal, or other codes. The installation, operation, monitoring, control, and maintenance of premises wiring systems supplying electricity to critical operations areas are addressed to ensure their survivability in the event of naturally ocurring hazards and human-caused events.

Article 720 Circuits and Equipment Operating at Less Than 50 Volts

DC or AC installations operating at less than 50 volts are covered by this article.

Article 725 Class 1, Class 2, and Class 3, Remote Control, Signaling, and Power Limited Circuits

Remote control, signaling, and power limited circuits not part of a device or appliance are covered by this article.

Article 727 Instrumentation Tray Cable: Type ITC

This article covers construction specifications, use, and installation of instrumentation tray cable, limited to applications at 150 volts or less and 5 amps or less.

Article 760 Fire Alarm Systems

All wiring, equipment, and circuits controlled by a fire alarm system are covered by this article (Figure 12–15).

Article 770 Optical Fiber Cables and Raceways

The installation of all fiber cables and raceways for fiber cables are covered by this article (Figure 12–16).

FIGURE 12–15 Fire alarm systems represent special conditions and are addressed in Chapter 7.

FIGURE 12–16 When optical fiber cable and raceways are installed, they represent a special condition and are addressed in Chapter 7.

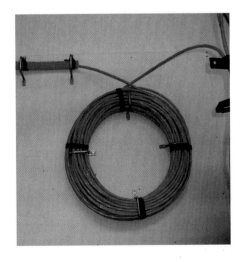

SUMMARY

Chapters 5, 6, and 7 are known as the "Special Chapters" due to the fact that they contain modifications and supplemental information and rules to add to the basic foundation of Chapters 1 through 4. These special chapters contain 62 articles that the user of the *NEC*® must become familiar with for proper application. When applying the rules of the *NEC*® to any electrical installation, it is imperative that you recognize and refer to the special chapters when special occupancies, special equipment, or special conditions are part of an installation.

REVIEW QUESTIONS

1. Do Chapters 5, 6, and 7 of the *NEC*® apply to all electrical installations covered by the *NEC*®?

2. Chapter 6 is subdivided into how many articles?

3. The scope of Chapter 5 of the *NEC*® is dedicated to Special _____ .

4. What part of which special article would address the installation of a temporary audio installation?

5. Chapter 7 of the *NEC*® addresses Special Conditions. What part of which article in Chapter 7 would apply if one were sizing overcurrent protection in a legally required standby system?

6. Do the definitions in Section 500.2 apply to Article 501?

7. Which five articles in Chapter 5 address entertainment venues?

8. How many parts is Article 690 subdivided into?

9. Does Part III of Article 725 apply to Class 1 circuits?

10. Which special article covers requirements for generation of power with fuel cells?

PRACTICE PROBLEMS

Using the Codeology Method

The following questions and steps to find the answer are designed to illustrate the Codeology method. While the steps to find your answer may seem lengthy, they are designed to illustrate the thought process to find the answer. The easiest way to make the Codeology method work is to silently "talk to yourself." Walk through these steps by silently talking to yourself and Codeology will be a natural response for quickly and accurately finding needed information in the *NEC*®.

Read the following questions and follow step by step using your codebook and the Codeology method.

1. In a Class I Division 1 location, a _____ branch circuit shall not be permitted.

Step #1
Qualify the question or need, look for key words
This question is about a Class I, Division 1 location.
Key words: *"Class I, Division 1 location"*
Think special occupancies

Step #2
Go to the table of contents, Chapter 5, the Special Occupancy chapter
Find the correct article   **501 Class I Locations**

Step #3
Further qualify your question or need and get in the right part of the article
Key words: *"Class I, Division 1 location, branch circuit"*
Look in **Part II Wiring**

Step #4
Read each section title until the correct section is found
Identify correct section or subdivision and find answer
Correct section/subdivision
Section   **501.40 Multiwire Branch Circuits**

2. In other than a single-family dwelling unit, an emergency switch is required for a spa. This switch must be within sight and not less than _____ feet from the spa.

Step #1
Qualify the question or need, look for key words
This question is about a spa.
Key words: *"required for a spa"*
Think special equipment

Step #2
Go to the table of contents, Chapter 6, the Special Equipment chapter
Find the correct article   **680 Swimming Pools, Fountains, and Similar Installations**

Step #3
Further qualify your question or need and get in the right part of the article
Key words: *"emergency switch is required for a spa"*
Look in **Part IV Spas and Hot Tubs**

Step #4
Read each section title until the correct section is found
Identify correct section or subdivision and find answer
Correct section/subdivision
Section   **680.41 Emergency Switch for Spas and Hot Tubs**

3. In the event of the failure of a normal source of power, an emergency system must be available within _____ seconds.

Step #1
Qualify the question or need, look for key words
This question is about an emergency system.
Key words: *"an emergency system must be available"*
Think special condition

Step #2

Go to the table of contents, Chapter 7, the Special Condition chapter

Find the correct article   **700 Emergency Systems**

Step #3

Further qualify your question or need and get in the right part of the article

Key words: *"failure of a normal source of power, an emergency system must be available"*

Look in **Part III Sources of Power**

Step #4

Read each section title until the correct section is found

Identify correct section or subdivision and find answer

Correct section/subdivision

Section   **700.12 General Requirements**

---

4. When a separately derived system operating at 120-volts line-to-line and 60-volts to grounding is serving sensitive electronic equipment, the maximum voltage drop on any branch circuit is _____ %.

Step #1

Qualify the question or need, look for key words

This question is about sensitive electronic equipment.

Key words: *"sensitive electronic equipment"*

Think special equipment

Step #2

Go to the table of contents, Chapter 6, the Special Equipment chapter

Find the correct article   **647 Sensitive Electronic Equipment**

Step #3

Further qualify your question or need and get in the right part of the article

Key words: *"sensitive electronic equipment, voltage drop"*

Article 647 is not separated into parts.

Step #4

Read each section title until the correct section is found

Identify correct section or subdivision and find answer

Correct section/subdivision

Section   **647.4 Wiring Methods**

First-Level Subdivision   **(D) Voltage Drop**

---

5. Power sources for power-limited fire alarm (PLFA) circuits shall not be supplied through ground-fault circuit interrupters or _____ circuit interrupters.

Step #1

Qualify the question or need, look for key words

This question is about a fire alarm system.

Key words: *"fire alarm (PLFA) circuits"*

Think special condition

Step #2

Go to the table of contents, Chapter 7, the Special Condition chapter

Find the correct article   **760 Fire Alarm Systems**

Step #3

Further qualify your question or need and get in the right part of the article

Key words: *"alarm (PLFA) circuits shall not be supplied through"*

Look in **Part III Power Limited Fire Alarm (PLFA) Circuits**

Step #4

Read each section title until the correct section is found

Identify correct section or subdivision and find answer

Correct section/subdivision

Section   **760.121 Power Sources for PLFA Circuits**

First Level Subdivision **(B) Branch Circuit**

6. Service equipment for a mobile home is not permitted to be mounted on or in the mobile home. The service equipment must be adjacent to the home and not more than _____ feet away.

Qualify the question or need, look for key words
> This question is about a mobile home.

> Key words: *"mobile home"*
> Think special occupancies

Go to the table of contents, Chapter 5, the Special Occupancy chapter
> Find the correct article   **550 Mobile Homes, Manufactured Homes, and Mobile Home Parks**

Further qualify your question or need and get in the right part of the article
> Key words: *"mobile home, location of service equipment"*

> Look in **Part III Services and Feeders**

Read each section title until the correct section is found
> Identify correct section or subdivision and find answer
> Correct section/subdivision
> Section   **550.32 Service Equipment**
> First-Level Subdivision   **(A) Mobile Home Service Equipment**

7. Color coding of intrinsically safe conductors is permitted only when they are colored _____ and there are no other conductors colored the same.

Qualify the question or need, look for key words
> This question is about an intrinsically safe system.
> Key words: *"intrinsically safe conductors"*
> Think special occupancies

Go to the table of contents, Chapter 5, the Special Occupancy chapter
> Find the correct article   **504 Intrinsically Safe Systems**

Further qualify your question or need and get in the right part of the article
> Key words: *"color coding of intrinsically safe conductors"*

> Article 504 is not separated into parts

Read each section title until the correct section is found
> Identify correct section or subdivision and find answer
> Correct section/subdivision
> Section   **504.80 Identification**
> First-Level Subdivision   **(C) Color Coding**

8. Storage batteries used in a photovoltaic system in a dwelling unit must have the cells connected to operate at less than _____ volts.

Qualify the question or need, look for key words
> This question is about a photovoltaic system.
> Key words: *"storage batteries used in a photovoltaic system"*
> Think special equipment

Go to the table of contents, Chapter 6, the Special Equipment chapter
> Find the correct article   **690 Solar Photovoltaic Systems**

Further qualify your question or need and get in the right part of the article
> Key words: *"storage batteries used in a photovoltaic system in a dwelling unit"*

> Look in **Part VIII Storage Batteries**

Step #4

Read each section title until the correct section is found

Identify correct section or subdivision and find answer

Correct section/subdivision

Section **690.71 Installation**

First-Level Subdivision **(B) Dwellings**

Second-Level Subdivision **(1) Operating Voltage**

9. A Class 1 power-limited circuit can only be supplied from a source of not more than 30 volts and _____ volt-amps.

Step #1

Qualify the question or need, look for key words

This question is about a Class 1 circuit.

Key words: *"Class 1 circuit"*

Think special condition

Step #2

Go to the table of contents, Chapter 7, the Special Condition chapter

Find the correct article **725 Class 1, Class 2, and Class 3 Remote-Control, Signaling, and Power-Limited Circuits**

Step #3

Further qualify your question or need and get in the right part of the article

Key words: *"Class 1 circuit, voltage and power limitations"*

Look in **Part II Class 1 Circuits**

Step #4

Read each section title until the correct section is found

Identify correct section or subdivision and find answer

Correct section/subdivision

Section **725.41 Class 1 Circuit Classifications and Power Source Requirements**

First-Level Subdivision **(A) Class 1 Power Limited Circuits**

10. All ballasts used in electric signs must be provided with _____ protection.

Step #1

Qualify the question or need, look for key words

This question is about an electric sign.

Key words: *"electric sign"*

Think special equipment

Step #2

Go to the table of contents, Chapter 6, the Special Equipment chapter

Find the correct article **600 Electric Signs and Outline Lighting**

Step #3

Further qualify your question or need and get in the right part of the article

Key words: *"ballasts used in electric signs, protection"*

Look in **Part I General**

Step #4

Read each section title until the correct section is found

Identify correct section or subdivision and find answer

Correct section/subdivision

Section **600.22 Ballasts**

First-Level Subdivision **(B) Thermal Protection**

11. General-purpose optical fiber cable raceways shall be listed as being _____ to the spread of fire.

Step #1

Qualify the question or need, look for key words

This question is about a fiber optic cable.

Key words: *"general-purpose optical fiber raceway"*

Think special condition

Step #2

Go to the table of contents, Chapter 7, the Special Condition chapter

Find the correct article **770 Optical Fiber Cables and Raceways**

Step #3

Further qualify your question or need and get in the right part of the article

    Key words: *"listed"*

    Look in **Part VI Listing Requirements**

Step #4

Read each section title until the correct section is found

    Identify correct section or subdivision and find answer

    Correct section/subdivision

    Section   **770.182 Optical Fiber Raceways**

    First-Level Subdivision   **(C) General-Purpose Optical Cable Raceway**

12. The disconnecting means for the connection to shore power in a marina is permitted to be a _____ or a switch.

Step #1

Qualify the question or need, look for key words

    This question is about a marina.

    Key words: *"shore power in a marina"*

    Think special occupancies

Step #2

Go to the table of contents, Chapter 5, the Special Occupancy chapter

    Find the correct article   **555 Marinas and Boatyards**

Step #3

Further qualify your question or need and get in the right part of the article

    Key words: *"disconnecting means, shore power in a marina"*

    Article 555 is not separated into parts.

Step #4

Read each section title until the correct section is found

    Identify correct section or subdivision and find answer

    Correct section/subdivision

    Section   **555.17 Disconnecting Means for Shore Power Connections**

    First-Level Subdivision   **(A) Type**

13. Emergency controls for circuits in gasoline dispensing equipment at an attended self-service gas station must be not more than _____ feet from the dispensers.

Step #1

Qualify the question or need, look for key words

    This question is about a gas station.

    Key words: *"gasoline dispensing equipment"*

    Think special occupancies

Step #2

Go to the table of contents, Chapter 5, the Special Occupancy chapter

    Find the correct article   **514 Motor Fuel Dispensing Facilities**

Step #3

Further qualify your question or need and get in the right part of the article

    Key words: *"emergency controls, gasoline dispensing equipment"*

    Article 514 is not separated into parts.

Step #4

Read each section title until the correct section is found

    Identify correct section or subdivision and find answer

    Correct section/subdivision

    Section   **514.11 Circuit Disconnects**

    First-Level Subdivision   **(B) Attended Self Service Motor Fuel Dispensing Facilities**

14. Autotransformer-type dimmers installed in a stage switchboard for theatre use must not exceed _____ volts between conductors.

Step #1

Qualify the question or need, look for key words

    This question is about a theatre.

    Key words: *"stage switchboard for theatre"*

    Think special occupancies

Step #2

Go to the table of contents, Chapter 5, the Special Occupancy chapter

Find the correct article   **520 Theaters, Audience Areas of Motion Picture and Television Studios, Performance Areas, and Similar Locations**

Step #3

Further qualify your question or need and get in the right part of the article

Key words: *"dimmers, stage switchboard for theatre"*

Look in **Part II Fixed Stage Switchboards**

Step #4

Read each section title until the correct section is found

Identify correct section or subdivision and find answer

Correct section/subdivision

Section   **520.25 Dimmers**

First-Level Subdivision   **(C) Autotransformer Type Dimmers**

---

15. Fuse enclosures for utilization equipment installed in Class II Division 2 locations must be _____ .

Step #1

Qualify the question or need, look for key words

This question is about a Class II location.

Key words: *"Class II Division 2"*

Think special occupancies

Step #2

Go to the table of contents, Chapter 5, the Special Occupancy chapter

Find the correct article   **502 Class II Locations**

Step #3

Further qualify your question or need and get in the right part of the article

Key words: *"utilization equipment installed in Class II Division 2"*

Look in **Part III Equipment**

Step #4

Read each section title until the correct section is found

Identify correct section or subdivision and find answer

Correct section/subdivision

Section   **502.135 Utilization Equipment**

First-Level Subdivision   **(B) Class II Division 2**

Second-Level Subdivision   **(3) Switches, Circuit Breakers and Fuses**

13

Chapter 8 of the *NEC®*, "COMMUNICATIONS SYSTEMS"

OUTLINE

OBJECTIVES

After completing this unit, you should be able to:
1. Associate the Codeology title for *NEC®* Chapter 8 as "Communications"
2. Identify the communications systems type of information and requirements contained in Chapter 8
3. Recognize key words and clues for locating answers in *NEC®* Chapter 8, the Communications chapter
4. Identify exact sections, subdivisions, list items, etc. to justify answers for all questions referring to Chapter 8 articles
5. Recognize that Chapter 8 numbering is the 800-series
6. Recognize, recall, and become familiar with articles contained in Chapter 8

OVERVIEW

The Codeology title for Chapter 8 of the *NEC®* is "Communications." Chapter 8 of the *NEC®* is an island; it stands alone from the rest of the *NEC®*. The arrangement of the *NEC®*, as required in Section 90.3, states that "Chapter 8 covers communications systems and is not subject to the requirements of Chapters 1 through 7 except where the requirements are specifically referenced in Chapter 8." There are four articles in Chapter 8, the 800-series. These four articles provide general requirements and information for all installations of communications systems.

NEC® CHAPTER 8 "COMMUNICATIONS SYSTEMS"

Chapters 1 through 7 of the *NEC* do not apply to any of the four Chapter 8 articles unless there is a specific reference in a Chapter 8 article to another area of the *NEC*. For example, in all four articles of Chapter 8, there is a specific reference to *Article 100 Definitions*. Each article in Chapter 8 recognizes all of the definitions in Article 100 as being applicable within that Article. Three of the articles include additional definitions which are applicable only within that article.

Article 800	800.2 recognizes all of Article 100 and adds eleven definitions which apply only within Article 800.
Article 810	810.2 recognizes all of Article 100. No additional definitions.
Article 820	820.2 recognizes all of Article 100 and adds six definitions which apply only within Article 820.
Article 830	830.2 recognizes all of Article 100 and adds seven definitions which apply only within Article 830.

In each of the four articles in Chapter 8, there is a section dedicated to referencing other articles; it is always the third section in each article (for example, 800.3). Other articles recognize requirements in Chapter 5, where hazardous locations are encountered, and 300.22(C) for permitted wiring methods in other space used for enviromental air. These examples are not all inclusive; there are multiple other references to requirements in Chapters 1 through 7 in the four Chapter 8 articles.

While the rest of the *NEC*® does not apply unless referenced, all four articles of Chapter 8 must be enforced when the *NEC*® is adopted. The *NEC*® includes signaling and communications conductors, equipment, and raceways. The *NEC*® is intended, as per Section 90.4, to give governmental bodies legal jurisdiction over electrical installations including signaling and communications systems. All four of the Chapter 8 articles cover specific types, methods, conductors, and equipment for communications systems as shown in Table 13–1.

TABLE 13–1 Layout of *NEC*® Chapter 8

NEC® Title:	Communications Systems
Codeology Title:	Communications
Chapter Scope:	Communications Systems Only

Article	Article Title
800	Communications Circuits
810	Radio and Television Equipment
820	Community Antenna Television and Radio Distribution Systems
830	Network Powered Broadband Communications Systems

DidYouKnow?

The scope of Article 800 is limited to communications circuits and equipment.

Article 800 Communications Circuits

Article 800 is separated into six parts to cover the installation of the following:

- Telephone installations
- Telegraph installations (except radio telegraph)
- Outside wiring for fire alarm and burglar alarm systems
- Telephone systems, equipment, installation, and maintenance (Figure 13–1)

FIGURE 13–1 Communications systems are covered in Chapter 8.

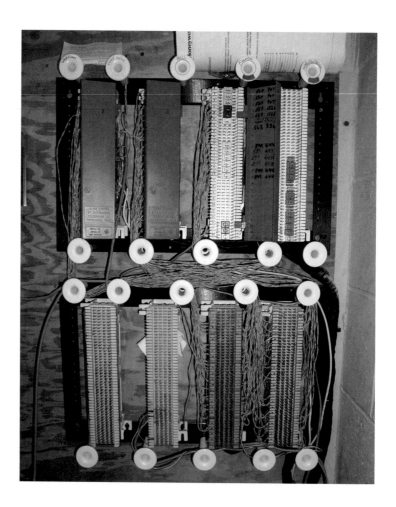

Article 800 is separated into six parts as follows:

Part I	General
Part II	Wires and Cables Outside and Entering Buildings
Part III	Protection
Part IV	Grounding Methods
Part V	Installation Methods Within Buildings
Part VI	Listing Requirements

Article 810 Radio and Television Equipment

Article 810 is separated into four parts to cover the installation of the following:

- Antenna systems for radio and television receiving equipment
- Amateur radio transmitting and receiving equipment
- Transmitter safety
- Antennas and associated wiring and cabling, including the following types:
  - Multielement antennas
  - Vertical rod antennas
  - Dish antennas (Figures 13–2a and b)

Article 810 is separated into four parts as follows:

Part I	General
Part II	Receiving Equipment—Antenna Systems
Part III	Amateur Transmitting and Receiving Stations—Antenna Systems
Part IV	Interior Installation—Transmitting Stations

Article 820 Community Antenna Television and Radio Distribution Systems

Article 820 is separated into six parts to cover the installation of the following:

- Cable distribution of radio frequency signals employed in community antenna television (CATV) systems (Figure 13–3)

Article 820 is separated into six parts as follows:

Part I	General
Part II	Coaxial Cables Outside and Entering Buildings
Part III	Protection
Part IV	Grounding Methods
Part V	Installation Methods Within Buildings
Part VI	Listing Requirements

Article 830 Network Powered Broadband Communications Systems

Article 830 is separated into six parts to cover the installation of the following:

- Network powered broadband communications systems that provide any combination of the following:
  - Voice systems
  - Audio systems
  - Video systems
  - Data systems
  - All interactive services through a network interface unit

FIGURE 13-2

Radio and television equipment are a type of communications system covered in Chapter 8.

(a)

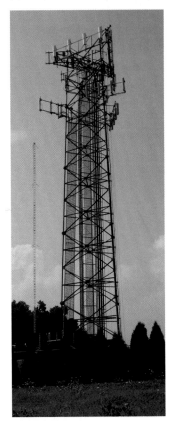

(b)

FIGURE 13-3 CATV installations are a type of communications system covered in Chapter 8.

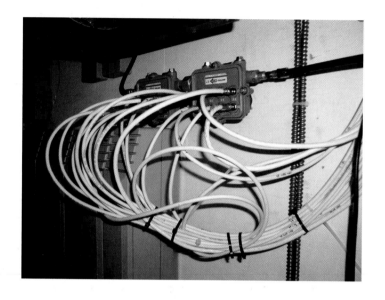

DidYouKnow?

The scope of Article 830 includes network-powered broadband communications systems that provide any combination of voice, audio, video, data, and interactive services through a network interface unit.

Article 830 is separated into six parts as follows:

Part I   General
Part II   Cables Outside and Entering Buildings
Part III   Protection
Part IV   Grounding Methods
Part V   Installation Methods Within Buildings
Part VI   Listing Requirements

SUMMARY

Chapter 8, which is dedicated to communications systems, stands alone from the rest of the *NEC®*. Chapters 1 through 7 apply only when there is a specific reference made in a Chapter 8 article. Chapter 8 consists of four articles as follows:

800   Communications Circuits

810   Radio and Television Equipment

820   Community Antenna Television and Radio Distribution Systems

830   Network Powered Broadband Communications Systems

These four Chapter 8 articles cover specific methods, conductors, and equipment for communications systems. While other areas of the *NEC®* may seem to be appropriately applied to a communications installation, such as conduit or box fill, Chapter 8 stands alone unless a specific reference is made to Chapters 1 through 7.

REVIEW QUESTIONS

1. Do Chapters 1 through 7 of the *NEC®* apply generally to all communications installations?

2. Chapter 8 is subdivided into how many articles?

3. The scope of Chapter 8 of the *NEC®* is dedicated to circuits and equipment for all types of _____ systems.

4. What part of which article in Chapter 8 would address the installation of cables inside of a building for cable TV?

5. Chapter 8 of the *NEC®* addresses communications systems. What part of which article would contain listing requirements for communications wire and cables?

6. When would a requirement located in Chapters 1 through 7 apply to an article located in Chapter 8?

7. What part of which article is referenced in 800.90(B)?

8. How many parts is Article 830 subdivided into?

9. Does Part II of Article 830 apply to grounding methods for network powered broadband communications systems?

10. Which article in Chapter 8 covers requirements for amateur radio antenna systems?

PRACTICE PROBLEMS

Using the Codeology Method

The following questions and steps to find the answer are designed to illustrate the Codeology method. While the steps to find your answer may seem lengthy, they are designed to illustrate the thought process to find the answer. The easiest way to make the Codeology method work is to silently "talk to yourself." Walk through these steps by silently talking to yourself and Codeology will be a natural response for quickly and accurately finding needed information in the NEC®.

Read the following questions and follow step by step using your codebook and the Codeology method.

1. Type _____ cable or type BL cable may be used as wiring methods for low-power network broadband communications systems as risers in space used for environmental air in dwelling units.

Step #1

Qualify the question or need, look for key words
This question is about network broadband communications.

Key words: "network broadband communications"

Think communications

Step #2

Go to the table of contents, Chapter 8, the Communications chapter.

Find the correct article   **830 Network Powered Broadband Communications Systems**

Step #3

Further qualify your question or need and get in the right part of the article

Key words: "wiring methods, low-power network broadband communications systems, riser air handling, in dwelling units"

Look in **Part V Installation Methods Within Buildings**

Step #4

Read each section title until the correct section is found

Identify correct section or subdivision and find answer

Correct section/subdivision

Section   **830.154 Low Power, Network Powered Broadband Communications System Wiring Methods**

First-Level Subdivision   **(B) Riser**

Second-Level Subdivision   **(3) One and Two Family Dwellings**

2. Radio transmitters installed indoors must have interlocks on all access doors to the transmitter, which disconnect all voltages of over _____ volts when any door is opened.

Step #1

Qualify the question or need, look for key words
This question is about radio transmitters.

Key words: "radio transmitters"

Think communications

Step #2

Go to the table of contents, Chapter 8, the Communications chapter.

Find the correct article   **810 Radio and Television Equipment**

Step #3

Further qualify your question or need and get in the right part of the article

Key words: "radio transmitters, installed indoors, interlocks, access doors"

Look in **Part IV Interior Installation—Transmitting Stations**

Step #4

Read each section title until the correct section is found

Identify correct section or subdivision and find answer

Correct section/subdivision

Section   **810.71 General**

First-Level Subdivision   **(C) Interlocks on Doors**

3. The minimum size grounding conductor for a telephone system installation is _____ AWG.

Qualify the question or need, look for key words

This question is about a telephone system installation.

Key words: *"telephone system installation"*

Think communications

Go to the table of contents, Chapter 8, the Communications chapter.

Find the correct article   **800 Communications Circuits**

Further qualify your question or need and get in the right part of the article

Key words: *"minimum size grounding conductor for a telephone system"*

Look in **Part IV Grounding Methods**

Read each section title until the correct section is found

Identify correct section or subdivision and find answer

Correct section/subdivision

Section   **800.100 Cable and Primary Protector Grounding**

First-Level Subdivision   **(A) Grounding Conductor**

Second-Level Subdivision **(3) Size**

4. Community antenna television (CATV) systems must be installed in a neat and _____ manner.

Qualify the question or need, look for key words

This question is about a community antenna television (CATV) system.

Key words: *"community antenna television (CATV)"*

Think communications

Go to the table of contents, Chapter 8, the Communications chapter.

Find the correct article   **820 Community Antenna Television and Radio Distribution Systems**

Further qualify your question or need and get in the right part of the article

Key words: *"installation of CATV system"*

Look in **Part I General**

Read each section title until the correct section is found

Identify correct section or subdivision and find answer

Correct section/subdivision

Section   **820.24 Mechanical Execution of Work**

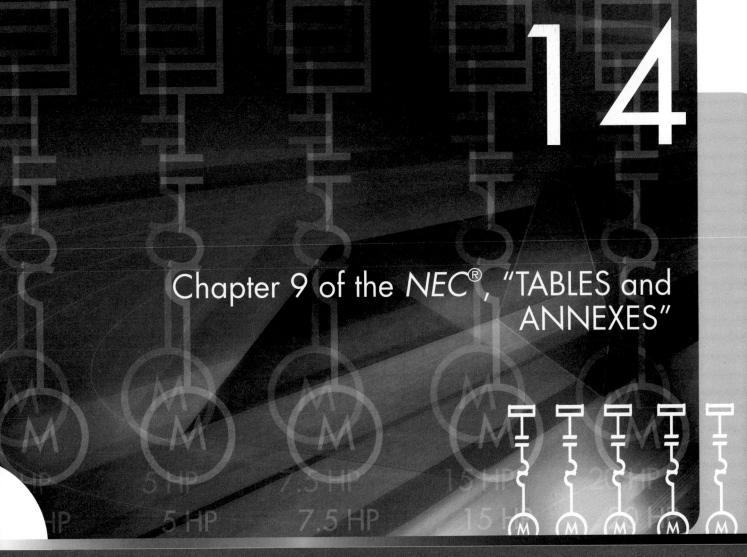

14

Chapter 9 of the *NEC®*, "TABLES and ANNEXES"

OUTLINE

OBJECTIVES

After completing this unit, you should be able to:

1. Associate the Codeology title for *NEC®* Chapter 9 as "Tables and Annexes"
2. Identify the specific type of information contained in the tables and annexes of Chapter 9
3. Recognize that tables in Chapter 9 apply as referenced elsewhere in the *NEC®*
4. Recognize that annexes are included in the *NEC®* for informational purposes only
5. Recognize key words and clues for locating references to Chapter 9, the Table and Annex chapter
6. Recognize, recall, and become familiar with tables and annexes contained in Chapter 9

OVERVIEW

Chapter 9 consists of 11 tables and eight annexes. Tables are applicable only when referenced in the *NEC®*. Annexes are for informational use only and are not an enforceable part of the *NEC®*.

TABLES

As required in Section 90.3, Chapter 9 contains tables that apply as referenced throughout the *NEC®*. These tables are extremely valuable tools for the user of the Code. A basic understanding of the types of tables in Chapter 9 and where they are referenced for use is necessary for quick reference to the correct table. Table 14–1 reviews the 11 tables in Chapter 9.

TABLE 14–1 *NEC®* Chapter 9 Tables

Table 1	Percent of Cross-Section of Conduit and Tubing for Conductors (Conduit Fill)
Table 2	Radius of Conduit and Tubing Bends
Table 4	Dimensions and Percent Area of Conduit and Tubing (Areas of Conduit or Tubing for the Combinations of Wires Permitted in Table 1, Chapter 9) (Conduit Fill, Tables for all circular raceways)
Table 5	Dimensions of Insulated Conductors and Fixture Wires
Table 5A	Compact Copper and Aluminum Building Wire Nominal Dimensions and Areas
Table 8	Conductor Properties
Table 9	Alternating-Current Resistance and Reactance for 600-Volt Cables, 3-Phase, 60 Hz, 75°C (167°F)—Three Single Conductors in Conduit
Table 11A	Class 2 and Class 3 Alternating Current Power Source Limitations
Table 11B	Class 2 and Class 3 Direct Current Power Source Limitations
Table 12A	PLFA Alternating Current Power Source Limitations
Table 12B	PLFA Direct Current Power Source Limitations

Table 1: Percent of Cross-Section of Conduit and Tubing for Conductors (Conduit Fill)

Table 1 is the benchmark for all permissible combinations of conductors in conduit and is commonly called the *conduit fill* table. Table 1 is short and to the point, as it gives only three types of conductor installations and permitted percentage of raceway fill. This table is referenced in the *NEC®* when raceway fill requirements exist. For example, using rigid metal conduit, Section 344.22 references the fill specified in Table 1, in Chapter 9. The nine notes to this table explain the application of these rules in all applications.

Table 2: Radius of Conduit and Tubing Bends

Table 2 provides a uniform minimum requirement for bends in conduit. This is necessary to prevent damage to raceways and to prevent the reduction of internal area for conductors. This table is referenced in the raceway articles. For example, using rigid metal conduit, Section 344.24 references the bend requirements of Table 2 in Chapter 9.

DidYouKnow?

The tables provided in Chapter 9 of the *NEC®* are applicable only when referenced elsewhere in the *NEC®*.

Table 4: Dimensions and Percent Area of Conduit and Tubing (Areas of Conduit or Tubing for the Combinations of Wires Permitted in Table 1, Chapter 9) *(Conduit Fill, Tables for all circular raceways)*

Table 4 is extremely useful for the user of the Code when determining conduit fill. Information provided by this table includes total internal area and permissible fill area for several applications. This table is referenced in note #6 to Table 1, making this table applicable whenever Table 1 is referenced in the *NEC*®.

Table 5: Dimensions of Insulated Conductors and Fixture Wires

Table 5 provides the dimensions of insulated conductors and fixture wires needed to determine permissible conduit fill. This table is to be used together with Table 4 to determine permissible combinations of conduit fill. This table is referenced in note #6 to Table 1, making this table applicable whenever Table 1 is referenced in the *NEC*®.

Table 5A: Compact Copper and Aluminum Building Wire Nominal Dimensions and Areas

Table 5A provides the dimensions of compact copper and aluminum building wire needed to determine permissible conduit fill. This table is to be used together with Table 4 to determine permissible combinations of conduit fill. This table is referenced in note #6 to Table 1, making this table applicable whenever Table 1 is referenced in the *NEC*®.

Table 8: Conductor Properties

This table provides conductor properties for all conductor sizes from 18 AWG to 2,000 kcmil. Information provided includes circular mil area for all AWG sizes, stranding, diameter, and DC resistance. Table 8 is referenced in note #8 to Table 1 for determining area for bare conductors, making this table applicable whenever Table 1 is referenced in the *NEC*®.

Table 9: Alternating-Current Resistance and Reactance for 600-Volt Cables, 3-Phase, 60 Hz, 75°C (167°F)— Three Single Conductors in Conduit

Table 9 provides resistance and impedance values necessary for determining proper conductor application when voltage drop or other calculations are required.

Table 11A: Class 2 and Class 3 Alternating Current Power Source Limitations and Table 11B: Class 2 and Class 3 Direct Current Power Source Limitations

Table 11(A) and Table 11(B) provide the required power source limitations for listed Class 2 and Class 3 power sources.

Table 12A: PLFA Alternating Current Power Source Limitations and Table 12B: PLFA Direct Current Power Source Limitations

Table 12(A) and Table 12(B) provide the required power source limitations for listed PLFA power sources.

ANNEXES

As required in section 90.3, the Annexes in Chapter 9 are not part of the requirements of the *NEC®* but are included for informational purposes only. A basic understanding of the types of annexes in Chapter 9 and the information they contain is necessary for the Code user to access this valuable information. Table 14–2 reviews the eight annexes contained in Chapter 9.

TABLE 14–2	Chapter 9, Annexes
Annex A	Product Safety Standards
Annex B	Application Information for Ampacity Calculation
Annex C	Conduit and Tubing Fill Tables for Conductors and Fixture Wires of the Same Size
Annex D	Examples
Annex E	Types of Construction
Annex F	Availability and Reliability for Critical Operations Power Systems; and Development and Implementation of Functional Performance Tests (FPTs) for Critical Operations Power Systems
Annex G	Supervisory Control and Data Acquisition (SCADA)
Annex H	Administration and Enforcement

Annex A: Product Safety Standards

Annex A provides a list of product safety standards used for product listing for products required to be listed in the *NEC®*.

Annex B: Application Information for Ampacity Calculation

Annex B provides application information for many different types of ampacity calculations, including conductors installed in electrical ducts.

Annex C: Conduit and Tubing Fill Tables for Conductors and Fixture Wires of the Same Size

Annex C is provided to aid the user of the Code in determining conduit fill when all conductors to be installed in a raceway are of the same size and type. This annex is informational only and is referenced in note #1 to Table 1.

Annex D: Examples

Annex D is provided to aid the user of the Code in making calculations required in the *NEC®*. The requirements for calculations are illustrated in the form of examples to aid the user of the Code in making similar calculations.

Annex E: Types of Construction

Annex E is provided to aid the user of the Code in the determination of the five different types of building construction.

Annex F: Availability and Reliability for Critical Operations Power Systems; and Development and Implementation of Functional Performance Tests (FPTs) for Critical Operations Power Systems

Annex F provides information for the Code user to assist in compliance with the commissioning requirements of Article 708 Critical Operations Power Systems.

Annex G: Supervisory Control and Data Acquisition (SCADA)

Annex G provides information useful for the implementation of a security control and data acquisition system that may be installed together with a critical operations power system described in Article 708.

Annex H: Administration and Enforcement

Annex G is provided as a model set of administration and enforcement requirements that could be adopted by a governmental body along with the electrical installation requirements of the *NEC®*.

SUMMARY

The tables contained in Chapter 9 of the *NEC*® apply only when they are referenced in the *NEC*®. These eleven tables are necessary to apply many provisions of the *NEC*®. Conduit fill calculations may include the use of several tables including the following:

Table 1: Percent of Cross-Section of Conduit and Tubing for Conductors *(Conduit Fill)*

Table 4: Dimensions and Percent Area of Conduit and Tubing (Areas of Conduit or Tubing for the Combinations of Wires Permitted in Table 1, Chapter 9) *(Conduit Fill, Tables for all circular raceways)*

Table 5: Dimensions of Insulated Conductors and Fixture Wires

Table 5A: Compact Aluminum Building Wire Nominal Dimensions and Areas

Table 8: Conductor Properties

Annexes are informational only and are provided to aid the user of the *NEC*®. These annexes are useful tools for the Code user providing useful information on product standards, ampacity calculations, conduit fill, examples, types of construction, cross-reference tables, the application of Article 708, and administration and enforcement.

REVIEW QUESTIONS

1. The tables located in Chapter 9 of the *National Electrical Code*® apply_____.
   a. at all times
   b. only in Chapters 5, 6, and 7
   c. whenever they are useful
   d. as referenced in the *NEC*®
2. Which table is referenced in raceway articles for conduit fill?
   a. Table 1
   b. Table 2
   c. Table 3
   d. Table 4

3. When the *NEC*® requires wiring methods, materials, and equipment to be listed, Annex _____ provides additional information on the product standards.
   a. G
   b. D
   c. F
   d. A
4. Annexes in Chapter 9 of the *NEC*® code are_____.
   a. informational only
   b. mandatory requirements
   c. applicable as referenced
   d. used only for special equipment

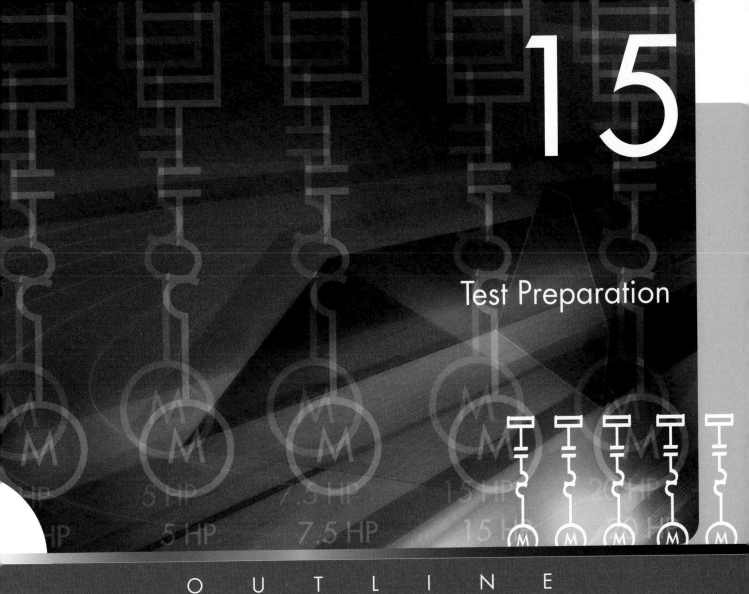

15

Test Preparation

OBJECTIVES

After completing this unit, you should be able to:

1. Recognize the importance of the Codeology method for persons taking *NEC®* exams
2. Identify the different types of exams
3. Identify the different types of exam questions
4. Recognize the importance of proper time management when taking an exam
5. Recognize the importance of developing a game plan before taking any exam

OVERVIEW

This unit is designed for those preparing to take an *NEC®* proficiency exam. The key to success in any type of exam is to be fully prepared. A Code exam can be quite different from other exams that you have taken in the past. Your ability to quickly and accurately find information in the *NEC®* using the Codeology method along with sound time management are the keys to success. In addition, a review of the different types of exams, questions, and testing methods must be understood for you to approach test day with an extremely high level of self-confidence.

ELECTRICAL CONTRACTOR, JOURNEYMAN/MASTER ELECTRICIAN, AND INSPECTOR EXAMS BASED ON THE *NEC*®

Electrical contractors, inspectors, and in many states, all electrical workers are required to be licensed. Preparing for an *NEC*® exam to become an electrical contractor, inspector, or an electrician requires many hours of study. There are three prerequisites that are necessary for all those about to take an *NEC*® exam: preparation, preparation, and preparation. The number one tool needed for success when taking an *NEC*® exam is a method to quickly and accurately find the needed information. Codeology is the most valuable tool to take with you on test day. Time management is the second most valuable tool.

Article-specific courses will be required for proficiency in areas such as grounding/bonding, calculations of services/feeders, calculations for box/conduit fill, calculations for conductor ampacity, and calculations for motors. All *NEC*® exams are broken down into groups of questions. The majority of questions in an *NEC*® exam will be answered without time-consuming calculations. For example, if an exam contained 80 questions total, 50 or more will not be time-consuming calculations. These are the questions that can be quickly and accurately answered using Codeology. The remaining questions are answered by doing the required calculations after locating the correct requirement using the Codeology method.

TYPES OF EXAMS

Written

Written exams usually consist of a test containing multiple-choice questions and an answer sheet. The answer sheets are usually numbered with four circles labeled, for example, as:

  #17   ① ② ③ ④

When an answer is determined, the applicant pencils in the appropriate answer as follows:

  #17   ① ② ● ④

When taking a written exam it is extremely important that each time an answer is placed on the answer sheet, the applicant place the answer in the proper location. For example, if an applicant skips question #2 on the test but not on the answer sheet, he/she could literally finish the exam with answers one question off on the answer sheet.

Computer

Computer-based exams will be a natural for the computer user. Applicants without basic computer skills should seek assistance and practice using a PC before test day. Most providers of *NEC*® exams will make sample questions available on your computer. Both applicants with and without basic computer skills should take advantage of an opportunity to become familiar with the computer test format. Exam providers will, in most cases, offer a short tutorial on the exam format for all applicants before the exam.

READING EACH QUESTION

Read each question and all of the multiple-choice answers in any exam before deciding to solve it quickly using Codeology, or highlight it to be solved later. It is essential that the question and all multiple choice answers be read because the key word or clue may be in one of the possible answers. By reading each question and the multiple-choice answers you will subconsciously retain those questions. As you move through the exam, other questions and/or multiple-choice answers will jog your brain to solve a skipped question.

TYPES OF QUESTIONS

Code questions can be broken down into two basic categories: those that can be solved quickly, which may contain a simple calculation, and those that will be time-consuming calculations problems.

General Knowledge Questions

General knowledge questions will be included in all *NEC*® Code exams. These questions are not designed to determine proficiency in the *NEC*® but to determine general knowledge of the electrical trade. General questions can include:

- Ohm's law
- Power formula
- AC formulas
- Inductance, capacitance
- Conversions (e.g., horsepower to watts)
- Voltage drop
- Local codes

Multiple-Choice Questions

There are typically four multiple-choice answers for an *NEC*® exam. When the answer is known, choose the correct option and move on to the next question. At the end of the exam all questions must be answered, unless there is no penalty for unanswered questions. For questions that are multiple-choice and are difficult to answer, try to eliminate one or more of the multiple-choice options. When more than one answer looks correct and "all of the above" is an option, it is probably the correct answer. When narrowed down to two answers from four possibilities and you are forced to guess, the odds of getting a correct answer are increased. Figure 15–1 shows preparation tips for successfully completing an *NEC*® exam.

TIME MANAGEMENT

When taking an *NEC*® exam there will be a given number of questions within a limited amount of time. For example, eighty questions could be given in a 4-hour period. Proper *time management* is essential when taking

DidYouKnow?

The two most essential tools which must be in place for a superior level of self-confidence on test day are:
- Codeology method
- Time-management skills

FIGURE 15-1 Exam tips.

AN *NEC*® EXAM GAME PLAN

■ PREPARATION

- Sharpen your Codeology skills
- Review all article-specific courses, including but not limited to:
  - Grounding and bonding
  - Box and conduit fill
  - Motors
  - Conductor ampacity
  - Calculations
- Review all general knowledge-type questions
  - Ohms law
  - AC formulas
  - Voltage drop
  - Control circuits
- Review all local codes that may be included

Exam (Game) Day

- Get a good night's rest
- **DO NOT CRAM** THE NIGHT BEFORE
  - Cramming will only increase test anxiety
- Eat a good breakfast
- Get to the exam location early and "relax"
- READ ALL OF THE EXAM INSTRUCTIONS BEFORE STARTING
- Read each question in order
- Highlight and skip all time-consuming calculations (do last)
- Do all questions that can be solved using CODEOLOGY
- Solve the calculations problems
  - Group them, do similar problems (e.g., motors) together
  - Solve them, easiest first, difficult last
- Answer all questions (most exams mark unanswered as incorrect)

any exam. On most exams, each question will have the same value or weight. For example, in an 80-question, 4-hour exam, each question would be worth 1.25 points. Each question would also be allotted only 3 minutes to solve. Time-consuming calculations make take 10 minutes or more to solve, while others using Codeology can be solved in 1 minute or less. When taking an *NEC*® exam, time-consuming calculations should always be skipped and solved last, even though the applicant may feel comfortable solving the problem. All of the questions are generally the same point value; therefore, it is a good strategy to solve all problems not involving time-consuming calculations first. After an *NEC*® exam, interviews of applicants would typically reveal the following two scenarios:

Scenario 1

Bob is taking a 4-hour, 80-question electrical contractor exam. Bob takes each question in order. The front of the test is heavily loaded with calculation-type questions. Bob is hung up on a few questions and his calculations do not match any of the multiple-choice answers. Bob checks his watch and realizes that 2 hours have passed and he is only on question 18. He has two hours left to complete the remaining 62 questions. Panic sets in and Bob starts getting nervous. His self-confidence has disappeared and he begins to believe that he will fail the exam. Bob's ability to calmly read and solve each question is severely compromised. Bob did not use his time wisely.

Scenario 2

Theresa is taking a 4-hour, 80-question electrical contractor exam. Theresa reads each question in order, highlighting time-consuming calculation-type questions to do later and using Codeology to quickly solve other questions. Theresa checks her watch and realizes that 2 hours have passed and she has completed 55 questions. She has 2 hours left to complete the remaining 25 questions. Theresa is pleased with her progress. Her self-confidence has increased and she feels that she will surely pass the exam. Theresa's ability to calmly read and solve each question is increased because of her renewed self-confidence. Theresa used her time wisely.

SUMMARY

Preparing for and taking any type of Code exam for an entry-level position or to become an electrical contractor can be an unnerving experience for many electrical workers. Without a game plan, a method to quickly solve basic questions, and time management skills, these exams are unnerving because the applicant is not properly prepared.

Proficiency in the *NEC*® will require more than Codeology and time management skills. It will also require article/topic-specific code courses to become competent in calculations, grounding, and other areas. However, on test day the two most important tools for the applicant are a method to quickly find needed information (Codeology) and time management.

Applicants must familiarize themselves with the type of test: paper or computer. Applicants must also understand the type of questions on an exam, the value of each question, number of questions, time allowed, and whether or not unanswered questions are marked as incorrect. Time-consuming calculations should be left to do at the end of the exam.

Proper preparation for an exam will result in a high level of self-confidence, allowing the applicant to be more relaxed. Skipping time-consuming problems early in the exam and using Codeology to solve all others will result in increased self-confidence. This will also allow ample time at the end of the exam to solve calculations in a self-confident and relaxed state.

REVIEW QUESTIONS

1. Code users taking a competency exam on the *National Electrical Code*® must take the following step/s to be completely prepared:
   a. Practice sound time management
   b. Prepare and study for all types of Code questions
   c. Be capable of quickly and accurately finding information
   d. All of the above

2. Using proper time management methods, the Codeology user will skip all time-consuming _____ problems and solve them after all other questions have been answered.
   a. Chapter 5
   b. calculations
   c. multiple-choice
   d. grounding

3. General knowledge-type questions on *NEC*® competency exams may include questions involving:
   a. Ohm's law and AC formulas
   b. Control circuits
   c. Voltage drop
   d. All of the above

4. Lack of proper preparation, including time management skills, the ability to quickly and accurately find information, and article-specific preparation will result in reduced _____ _____ and ultimately failure on an exam.
   a. self-confidence
   b. ability to guess
   c. good luck
   d. speed reading

American National Standards Institute (ANSI) A private, non-profit organization that administers and coordinates the United States' voluntary standardization system. ANSI enhances the standard-making process by promoting and facilitating voluntary consensus standards and safeguarding their integrity. The *NEC®* is an ANSI standard.

Annex As required in Section 90.3, annexes are not part of the requirements of the *NEC®* but are included for informational purposes only.

Article Articles are major subdivisions of chapters in the *NEC®*.

BUILD "BUILD" is the Codeology title for "Chapter 3 Wiring Methods and Materials" of the *NEC®*. All electrical installations must be built after they have been properly planned.

Chapter Chapters are major subdivisions of the *NEC®*. There are nine chapters and the Introduction in the *NEC®*.

Code arrangement The arrangement of the *NEC®* is as required in Section 90.3. Chapters 1 through 4 apply generally to all electrical installations. Chapters 5, 6, and 7 are special chapters and supplement or modify Chapters 1 through 4. Chapter 8 covers communications and is not subject to the provisions of Chapters 1 through 7 unless there is a specific reference made in a Chapter 8 article. Chapter 9 contains tables that apply only when referenced elsewhere in the *NEC®*. Annexes are informational only and are not an enforceable part of the *NEC®*.

Code-Making Panel Also known as the Technical Committee (TC), it is made up of volunteers; each volunteer has a committee classification. The Technical Correlating Committee, the Standards Council, and the panel chairman work diligently to ensure each Code-Making Panel (CMP) is properly balanced. The *NEC®* has 19 Code-Making Panels; each panel is charged with specific articles within their purview and panel makeup.

Codeology A method that allows the Code user to find needed information accurately and quickly in the *NEC®*. This Codeology method results in a superior confidence level in one's capability to find pertinent information in the *NEC®* very quickly.

Comments Comments are formally submitted to NFPA by any interested party. Comments are generated after the ROP stage. Each comment is directed toward a specific proposal. Comments can suggest continued acceptance, rejection, or a supplement or modification to the proposal, provided new material is not introduced.

Committee membership classifications The committee list and the scope for each Code-Making Panel is in the front of the *NEC®*. Each Technical Committee member has an identification letter/s listed with their name and employer. This identification is in brackets, such as [L] for Labor. The committee classification is part of the balancing process of each panel. Most organizations represented have a principal member and an alternate.

COMMUNICATIONS "COMMUNICATIONS" is the Codeology title for "Chapter 8 Communications Systems" of the *NEC®*. Chapter 8 is dedicated to communications systems and is not subject to the provisions of Chapters 1 through 7 unless a specific reference is made in a Chapter 8 article.

Consensus standard A standard, such as the *NEC®*, that is revised through a consensus process; all ANSI standards are consensus standards. The consensus process consists of a system by which all interested persons and organizations may submit proposals and comments to change the document during the revision cycle. The consensus process also includes balanced committees made up of volunteers who must come to a consensus of opinion, or general agreement on all proposed changes to the standard.

Cross-reference tables There are six cross-reference tables included in the *NEC®* to aid the user of the Code. These tables are provided when there are multiple modifications or supplemental requirements related to the scope of an article located elsewhere in the *NEC®*.

Definitions When necessary for proper application of the *NEC®*, terms are defined. Terms used in more than one article are located in Article 100. Terms defined that are used in a single article are located in the second section of that article.

Diagrams The *NEC®* includes diagrams to illustrate the application of an article, section, or provide basic information in ladder-type diagrams to aid the Code user.

Drawings The *NEC®* includes drawings to illustrate the application of an article, section, or provide basic information in drawing form to aid the Code user.

Exceptions Exceptions are only used in the *NEC®* when they are absolutely necessary. Exceptions are always italicized in the *NEC®* for quick and easy identification.

Explanatory text Explanatory text is included in the *NEC®* in the form of Fine Print Notes. This material is included to aid the user of the Code through additional information, examples design considerations and other Code references.

Extract material The *National Electrical Code®*, also known as NFPA-70, is one of many documents published by the National Fire Protection Agency. Where another NFPA document has primary jurisdiction over material to be included in the *NEC®*, the material is extracted into the *NEC®*.

Fine Print Note Fine Print Notes (FPN) are explanatory material and are not an enforceable part of the *NEC®*. FPNs include but are not limited to the following types: informational, reference (referencing other sections or areas of the Code or other codes) design, suggestions, and examples.

GENERAL "GENERAL" is the Codeology title for "Chapter 1 General" of the *NEC®*. All electrical installations are subject to general rules that apply in all electrical installations. Chapter 1 of the *NEC®* contains these general rules.

List item When necessary, list items are included in sections, subdivisions, or exceptions to fully illustrate the intended requirement.

Mandatory language The use of the terms "shall" or "shall not" in the *NEC®* represent mandatory requirements.

National Electrical Code® (NEC) NFPA-70, the installation document used daily by electricians, wiremen, engineers, and inspectors.

National Fire Protection Association (NFPA) A nonprofit organization that works toward reducing the burden of fire and other hazards (including electrical hazards) on the quality of life by advocating scientifically based consensus codes, standards, research, training, and education.

NEC® Style Manual The rules that govern the structure of the *NEC®* are known as the "*National Electrical Code®* Style Manual."

Outlines The *NEC®* includes outlines to illustrate the application of an article, section, or provide basic information in outline form to aid the Code user.

Part Parts are major subdivisions of articles in the *NEC®*.

Permissive text The use of the terms "shall be permitted" or "shall not be required" in the *NEC®* represent permissive language.

PLAN "PLAN" is the Codeology title for "Chapter 2 Wiring and Protection" of the *NEC®*. All electrical installations must be planned to suit many needs. These needs include but are not limited to the design of an electrical installation as well as compliance with all applicable codes and standards.

Proposals Proposals are formally submitted to NFPA by any interested party. Proposals are generated and submitted to NFPA and are the first step in the 3-year revision cycle. Proposals may suggest changes, modifications, or deletions of material in the *NEC®*.

Report on Comments (ROC) The ROC is a document produced by committee members after deliberating and acting on all comments submitted. This document is an important part of the comment or ROC stage of the 3-year cycle to revise the *NEC®*. All interested persons or organizations that have submitted a comment directed to an individual proposal generated in the ROP stage are sent a copy of the ROC. The ROC

is also made available to the public. The document known as the ROC becomes part of the permanent record documenting the continuing evolution of the *NEC®*.

Report on Proposals (ROP) The ROP is a document produced during the committee meetings to deliberate and act on all proposals submitted. This is known as the proposal or ROP stage of the three-year cycle to revise the *NEC®*. All interested persons or organizations submitting a proposal are sent a copy of the ROP. The ROP is also made available to the public. The document known as the ROP becomes part of the permanent record documenting the continuing evolution of the *NEC®*.

Section Each part of an article is subdivided into separate sections, each of which is dedicated to a separate rule under the title of the specific part. These logical separations of requirements into separate sections are individually titled to identify the rules contained and are designed to aid the Code user to quickly find the necessary information.

Service point The point of connection between the facilities of the serving utility and the premises wiring.

SPECIALS Chapters 5, 6, and 7 are known as the Special chapters. The information and requirements included in these Special chapters will supplement or modify the information and requirements of Chapters 1 through 4.

Standards Council The Standards Council is composed of 13 volunteers appointed by the Board of Directors. The Standards Council oversees all activities on the development and revision of NFPA codes and standards. The primary function of the council is to assure that due process and fairness are upheld throughout the development and revision of NFPA codes and standards.

Subdivision Sections are sometimes subdivided to logically present the given rule or rules. These are called subdivisions. Sections are sometimes divided into as many as three levels of subdivisions to clarify a requirement.

Table of contents In the front of the *NEC®*, the table of contents provides the Code user with the outline form of the *NEC®*. Using the Codeology method, the table of contents is the starting point for all inquiries into the *NEC®*.

Tables As required in Section 90.3, Chapter 9 contains tables which apply as referenced throughout the *NEC®*.

Technical Comments (TC) Also known as the Code-Making Panel, it is made up of volunteers, each with a committee classification. The Technical Correlating Committee, the Standards Council, and the panel chairman work diligently to ensure each technical committee is properly balanced. The *NEC®* has 19 technical committees, each of which is charged with specific articles within their purview and panel makeup.

Technical Correlating Committee (TCC) Also known as the TCC, it is made up of volunteers, each with a committee classification. The Technical Correlating Committee oversees the actions of the nineteen Technical Committees or Code-Making Panels throughout the 3-year revision cycle of the *NEC®*.

Terms Language or mode of expression used in the *NEC*®. Terms defined in the *NEC*® include single words such as "ampacity" and phrases such as "voltage to ground."

Units of measurement The *NEC*® uses both the SI system (metric units) and inch-pound units. The SI system is always shown first and the inch-pound are second and placed in parentheses.

USE "USE" is the Codeology title for "Chapter 4 Equipment for General Use" of the *NEC*®. Chapter 4 of the *NEC*® is dedicated to electrical equipment and necessary components designed to consume electrical energy.